PACIFIC SALMON

PACIFIC SALMON

& STEELHEAD TROUT

R.J. Childerhose
Marj Trim

University of Washington Press
SEATTLE

Copyright© Minister of Supply and Services Canada, 1979
FS34-1979/1
Published in co-operation with Department of Fisheries and Oceans Canada (Pacific) and the Canadian Government Publishing Centre, Department of Supply and Services, Ottawa, Canada

Published in the United States of America by University of Washington Press, 1979 by arrangement with Douglas & McIntyre Ltd., North Vancouver.

Library of Congress Cataloging in Publication Data
Childerhose, R J
　Pacific salmon.

　1. Pacific salmon. 2. Steelhead (Fish) 3. Fishery conservation—Northwest, Pacific. I. Trim, Marj, joint author. II. Title.
QL638.S2C43　597'.55　78-65830
ISBN 0-295-95642-9

All rights reserved. No part of this book may be reproduced or transmitted in any form or by any means, electronic or mechanical, including photocopy, recording, or any information storage or retrieval system, without permission in writing from the publisher.

Design by Bev Leech
Jacket design by Nancy Legue-Grout

Species paintings by Harry Heine
Drawings by Joey Morgan

Typesetting by Domino-Link Word and Data Processing Ltd.
Color separations and printing by Herzig Somerville
Bound by the T. H. Best Company

To J. Ronald MacLeod
WHOSE EFFORTS AND DETERMINATION RESULTED IN
THE SALMONID ENHANCEMENT PROGRAM

Acknowledgements

This book is based on the works and writings of scientists and managers of the Pacific Region of Fisheries and Oceans Canada and the University of British Columbia.

In particular:

D.F. Alderdice	J.R. MacLeod
R.A. Bams	J.H. Mundie
J.R. Brett	F.K. Sandercock
C. Groot	J. Sibert
P.A. Larkin (U.B.C.)	J.G. Stockner
R.J. LeBrasseur	F.C. Withler
D. MacKinnon	F.E.A. Wood

Contents

Preface 1

History 5

Biology 31
SPAWNING 33
FERTILIZATION 35
INCUBATION 36
HATCHING / ALEVIN 37
FRY 39
FRY MIGRANTS 40
ESTUARINE LIFE / OCEAN DISTRIBUTION 43
HOMING 44
FISHERY 45

Environment 49
STREAMS 50
ENEMIES 51
FOOD 52
DAMS 53
POLLUTION 55
HOMECOMING HAZARDS 74
HELL'S GATE 76

Color Plates and Maps

Sockeye salmon 13
Pink salmon 15
Coho salmon 17
Chinook salmon 19
Chum salmon 21
Steelhead trout 23
The Catch and the Environment 57
Hazards 79
Spawning Channels 95
Hatcheries 109
Ocean Distribution of B.C. Salmon Species 141
Location of Some B.C. Enhancement Projects 143
Inshore Fishery Management 144
The Salmon Run 145

Enhancement 89

GENETIC RESEARCH 91

SPAWNING CHANNELS 93

INCUBATION UNITS 105

HATCHERIES 106

TRANSPLANTS 108

HATCHERY OPERATIONS 121

HATCHERY LIMITATIONS 123

FISH MARKING 124

MANAGEMENT 127

REGULATION / MIXED STOCKS 128

WATERSHED CONSERVATION 129

LAKE ENRICHMENT 130

STREAM ENGINEERING 134

FISHWAYS 135

AQUACULTURE 138

ENHANCEMENT 139

Afterword 157

Preface

"The business of America is business," may be a cliché, but it is true. The history of the white man in North America has been an unceasing quest for wealth. The wealth is in natural resources—land, trees, minerals, fish—and to turn these into money we have encouraged industry to apply its technology with ruthless efficiency.

Times and ideas change. We realize now that mankind's survival depends on the husbanding of the finite supply of resources on earth. What was once taken for granted is now challenged; no longer can industry concern itself only with profits. Controversies flare around new projects such as the damming of wild rivers to provide electricity for energy-hungry people. Clear-cutting of forests or strip mining evoke public wrath. In the resulting arguments the phrase "wise use of resources" always comes up. But who decides what is "wise"? How do we know what is in the "public interest"? When there are two or more conflicting public interests, who decides which will be served?

"Wisdom" in resource development requires knowledge not only of the resource itself but also of the ecosystem of which it is a part. It means understanding society's demands—economic, social, political—as well as the limits of technology. Any management decision affecting a natural resource must be carefully examined.

An example is our Pacific salmon and steelhead trout resource. The consequences of man's activities over the past 100 years on the salmonid species (which include the sea-run trouts such as steelhead and cutthroat) have been decimation of fish stocks and the destruction of their habitat. The decline in the abundance of salmon began in the gold rush days of the last century, was greatest in the first two decades of the present century, and continued into the 1960s.

In the early years of the commercial fishery, the industry concentrated on the most valuable species of salmon in the most accessible areas. When one species or one area declined in productivity, the fishing fleet would shift to another species or another area. It was not until the late 1930s that the industry had seriously fished all species in all areas; by then, major declines in many stocks had occurred.

At the end of World War II the Pacific area "boomed" with industrial and urban growth. Degradation of the environment, overfishing, illegal fishing, and the effects of inadequate fisheries management knowledge added to the pressures on already depressed stocks. The Canadian government responded to the crisis by providing more protection and instituting biological studies to learn more about the management of fisheries. By the end of the 1960s the decline in British Columbia salmon was halted. Since then the runs have stabilized at levels about one half of what they were before 1900.

Although the decline of salmon stocks appears to have been stopped because of improved protection and management, whether salmon will survive remains in doubt. Losses are still occurring as a result of man-made changes in the watersheds and estuaries. The threats to the salmon from logging, forest fires, road building, pulp mills, mining, power dams, port development, and sewage disposal are easy to see. Other forms of damage, such as chemical and thermal pollution, are insidious, causing fractional but cumulative losses. The consequence is that most gains from improved fisheries management over the past two decades have been nullified by losses through debasement of habitat.

Another negative factor is that Canada's Pacific salmon are vulnerable to capture by foreign fishing fleets at sea and as the fish migrate through the coastal areas of the United States. The same is true for American salmon passing through Canadian waters. This factor has been mitigated to some degree by international treaties and the 200-mile coastal fishing limit. But the salmon stocks fished in Canadian and American waters as they return to spawn in their home streams remain largely unprotected. Bilateral agreements restrict salmon fishing by fishermen of both countries in certain areas of their coastal waters; nevertheless, large numbers of salmon originating in one country are taken by the fishermen of the other. Canadians may catch between two to three million American salmon each year, and American fishermen take a like number of bred-in-Canada salmon. With the exception of the Fraser River Salmon Convention of 1937, which provided for equal sharing of the cost of enhancement for sockeye and pink salmon in the Fraser River system and an equal split of the catches of these species in the Convention area, there is no mechanism for restricting the catch of each other's salmon. Since both countries have the capacity to intercept the other's salmon, it is not in the long-term interest of either to increase the rate of interceptions but rather to agree to some measures for control.

On the positive side, we are better able to forecast the size of returning runs and this in turn leads to more precise regulation of their harvesting. Each year's experience further improves the management capability.

Progress is being made with techniques developed over the years to improve the survival of the salmon during the freshwater stages of their life cycle. Starting in 1885 with the first hatchery, the Canadian government intensified the effort after World War II by building fishways, spawning channels, and flow control systems in British Columbia. Much effort, too, has been devoted to the protection and improvement of natural stocks and natural spawning areas. The goal has been to ensure that inbound salmon escaping the fishermen's nets and trolls can safely reach their spawning areas and that these areas are clean and ready for their arrival.

The success of these initial efforts has led to a 20-year program of salmonid enhancement which, primarily by means of artificial propagation, will double present-day stocks and restore them to their pre-1900 levels.

Although the salmon resource is administered by the federal government, the British Columbia government has a substantial interest in the program. The province may be asked to co-operate in multiple use of land and freshwater resources to make them available for salmonid enhancement. Also, the program will stimulate development in the rural coastal zone, a matter of considerable interest to the province. At the same time, production of steelhead trout will have to be increased, for they would be fished along with the enhanced stocks of salmon, and their smaller numbers would be decimated, particularly if heavier fishing were permitted.

This scheme to restore the great salmon runs will create new wealth for Canada. Fishermen's incomes will increase, as will revenue in both the primary and processing sectors of the commercial fishery. Manufacturers in other parts of the country will benefit as more money becomes available for consumer goods. Canada's balance of payments will improve through increased exports of salmon products. Improved recreational fishing will stimulate tourism and the service industries which cater to sports fishermen. British Columbia's growing Indian population will have more salmon for their traditional uses.

Twenty years ago the problem was to find markets for Pacific coast fish; now the problem is to find enough fish to meet market demands. Ironically, foreign nations have been more alert than we to the potential of stocks off our shores. European and Asian fleets cruise the oceans of the world, bringing the stocks of fish (and whales) under pressures so great that their existence has been threatened. One by one the fish stocks of the world have been subjected to overfishing until one by one they have declined.

A case in point is that of the Atlantic salmon. By the early 1970s that great silvery resource had all but been destroyed. Danes, Norwegians, and

Canadians fished them in the Davis Strait; Newfoundlanders set drift nets offshore and barricades of traps in the rivers. In 1972 the federal government restricted commercial fishing of the Atlantic salmon, and the endangered fish began to recover. Now the salmon runs are threatened in the rivers of Quebec, New Brunswick, and Newfoundland by poachers.

The problem of conservation remains: it is always easier to destroy than to create. The fight to save the salmon, the forest, the river, must be unending, for once a resource is destroyed, it is gone.

We are lucky that the remaining stocks of Pacific salmon provide a healthy base from which to start. With time, our knowledge of salmonid enhancement principles will grow. Old problems will be resolved. New technology will emerge. The problem of world hunger will not be solved by Canada's Salmonid Enhancement Program, but if by management we can improve our salmon harvest, we will have demonstrated that the oceans can still be cultivated to produce high-quality protein.

History

Long before the coming of the white man, the natives of the north Pacific coast called themselves "The Salmon People." Although they ate shellfish, bear, venison, and berries in season, salmon was the staple of their life. This was true for the Indians of interior British Columbia as well. To the west coast Indians the salmon was as the buffalo was to the plains Indians, as the reindeer was to the Inuit.

The aboriginals were animists. They believed that every animal, tree, or brook had a guardian spirit and that it was necessary to placate this spirit before killing the animal, cutting the tree down, or damming the stream with a weir. Every tribe that knew the fish had a version of the Kwakiutl's "First Salmon Ceremony."

O Supernatural Ones, O Swimmers, I thank you that you are willing to come to us.
Don't let your coming be bad, for you come to be food for us.
I beg you to protect me and the one who takes mercy on me, that we may not die without cause, Swimmers.

Similar rituals were held for the immortal souls of the deer or bear and before picking the first berry crops.

The white man came, derided the Indian's primitive religion, and, having no such regard for other forms of life, destroyed them for profit. Then, as now, the white race felt free to exploit nature.

By the seventeenth century the northern Europeans had developed technological skills that had already transformed their portion of the world. The iron plow converted vast tracts of wild land into ribbon-like farms. Windmills pumped water for irrigation. Waterwheels provided power for ripping timbers, milling grain, and pumping the bellows of blast furnaces.

The ecological penalties went unnoticed. Lush hills and valleys became black deserts whose dunes were coal mine wastes. By the year 1285 London had a smog problem, brought on by the thousands of soft coal fires used for heating. Entire forests went under the ax for the building of wooden ships. Iron ore was mined for cannon, potash and sulfur for gunpowder. And since gunpowder also required charcoal, more forests were leveled.

Technological superiority—notably the fleets of wooden ships with cannon—allowed the Europeans to spill out over the rest of the world, conquering, looting, and colonizing. The implacable Spaniards conquered for "glory, gold and God." The English fur traders who arrived on the west coast of North America were less bloody-minded than the Spaniards, but even for them, nature and the natives were there to be conquered. Pagan Indians might believe that every gushing spring, every stream, every creature had its own spirit; Christians *knew* that it was God's will that man use nature for man's proper ends.

When Alexander Mackenzie walked out of the mountains and the bush to the Pacific in 1793—becoming the first white to traverse the continent north of Mexico—he reached the coast at Bella Coola, site of a large native settlement. The Indians, hospitable though they were, objected to his iron kettle and took it away from him because its smell would drive the salmon from the river. Similarly, when he went to leave, Mackenzie recorded that their chief objected to his "embarking venison in a canoe on their river, as the fish would instantly smell it and abandon him [the chief], so that he, his friends and relations, must starve."

Mackenzie had been dependent for food on salmon dried by the natives of the interior. Simon Fraser, descending the great river that bears his name, relied on dried salmon for the entire journey. Extracts from Simon Fraser's diary, 1808, read:

Sunday, May 29—Cold morning. We were underway at 4. Went ashore upon an island and secured a bale of salmon for our return.

Later that day—Here we put three bales of salmon into cache and carried the rest through very rugged country.

Monday, May 30—At 6 we put to shore at a large house; found a cache of fish. After taking a few salmon and leaving the value we secured the rest for the owners . . . A little below we put ashore again, and left a bale of salmon in cache.

Tuesday, May 31—The chief and the Indians, recommended to our attention yesterday, presented us with dried salmon and different kinds of roots.

Geographer-explorer David Thompson was equally dependent on dried salmon. Thompson, after five years of searching for the headwaters of the Columbia River (from 1806 to 1811), finally found it and followed it to the sea. His goal—as had been Simon Fraser's—was to discover a direct water route from the mountains to the mouth of the Columbia and, once there, to establish a fur trading post for his employers, the North West Company. He was too late. John Jacob Astor had arrived there three months earlier and had built Fort Astoria for his own Pacific Fur Company.

Furs brought the early Europeans to the western mountain country, but trade was dependent on the salmon. Like the natives, the white traders wintered on dried salmon. If a salmon run failed in a watershed, everyone in that area faced starvation.

Bales of preserved salmon were used for barter by the inland tribes. The salmon flesh was dried, either in the sun or in smoke houses, then was pounded between stones. The resulting pulp was tightly packed into woven baskets made from grass, rushes, or cedar strips and waterproofed with a lining of stretched salmon skins. These moisture-tight packages were about one foot in diameter by two feet long and weighed between 70 and 100 pounds. Their size and shape were ideal for canoe transport. Stored in a dry place, the salmon would keep for years.

Usually the Indians preserved their salmon by sun drying. In rainy areas smoke houses were needed, and in some places smudge fires were used to discourage flies. Sometimes the salmon was lightly smoked for flavor before going on the racks for curing by sun and air. The process took from two to three weeks.

It is described in the 1827 report on Fort Alexandria by Chief Trader Joseph McGillivray:

> The salmon when cured loses 4/5 of its weight, becomes crisp—and of reddish appearance—rendered blackish by the incessant smoke which must daily be kept up to prevent the flies from settling on it . . . Fish thus cured forms their constant diet without condiment—and what is a remarkable fact—that it actually files the teeth to the very gums. I have observed many young men about 25 years of age who had their teeth half worn away—and at 40 they have positively none. The same effect operates on the whites. Its ravages are not so perceptible as we come to these countries at an advanced period in life.

That same year, 1827, the salmon run failed on the upper reaches of the Fraser, and everyone from Fort Kamloops to Fort St. James went hungry. By now the Indians were dependent on the trading posts for food, blankets, clothing, traps, guns, and ammunition. In the summer a heedless Indian might trade his salmon for tobacco and tea, knives and needles, buttons and beads. And rum. In winter, reduced to destitution, he would appear at the fur post begging for food.

Since the whole industry depended on the hunting skills of the Indian, the trader would find himself despatching dogsled teams to other forts for emergency supplies of salmon.

Lake Babine, northwest of Fort St. James, is the main spawning ground for Skeena River salmon. When white fur traders arrived there in 1812, the Babine Indians had never known a poor salmon season. Accordingly, in 1822 a fort was established for the purpose of supplying the New Caledonia district with dried salmon.

But the fur trade was already collapsing. In the three decades from 1795 to 1825 intensive trapping had exhausted whole regions. Interior British

Columbia was one of them. David Thompson had such success in the winter of 1811-12 trading for furs at what is now Kamloops that his company hastened to build a fort there. By 1828 the trade at Fort Kamloops had so dwindled that it made no profit.

Steel traps and firearms had increased the Indian trapper's efficiency to the point where populations of small fur-bearing animals were decimated. It was man versus nature. The fur industry began the despoliation of natural resources that has characterized the industrialization of North America.

The first trader to consider salmon as a replacement commodity for declining furs was Archibald McDonald. He was in charge of Fort Langley, 35 miles from the mouth of the Fraser. By trial and error he succeeded in preserving salmon fillets between layers of salt. In the 1830s the Hudson's Bay Company began a trade in salmon, buying from the Indians and shipping the salt-cured fish in barrels to Hawaii, Asia, and elsewhere.

The old fur brigades—horse caravans crawling the river canyon trails of interior British Columbia—became salmon brigades. Fort Babine supplied a specialty: Skeena River sockeye. As late as 1846 there is a record of 30,000 salmon sent out from Fort Babine in one season. The salmon stocks never failed in Babine Lake, but by the turn of the century the Indian population there had shrunk from 2,000 to 250. The Indians declined throughout British Columbia during the 1800s, dying by the thousands from smallpox, measles, tuberculosis, influenza, venereal disease, and alcoholism.

The fur trade had needed the Indians, first to trap, prepare, and transport the furs to the trading post; secondly to guide, paddle, and pack them over the long brigade trails from the interior to the seaport. When the fur trade gave way to mining and the whites began arriving in large numbers, the Indians were no longer needed.

Coal was discovered and mined on Vancouver Island in 1835 and then in the Queen Charlottes. Lode gold and copper deposits were worked on Vancouver Island and the Gulf Islands. But it took the discovery of placer gold on the Fraser to spark the gold rush of 1858. Hordes of miners, fresh from the disappointments of the earlier gold and silver rushes of California and Nevada, jammed the lower reaches of the Fraser and Thompson rivers. They slept under canvas in placer camps and lived on a thrice-per-day-diet of salmon.

Scarcely any of them made money. When the gold in the sands of the Fraser gave out, many returned to California. Others kept following the gold strikes north. There was the Cariboo rush of the 1860s followed by the Cassiar strike in the 1870s. Each find was quickly exhausted; in 1867 the output of British Columbia gold mines was nearly $4 million, but in 1871—the year British Columbia joined the rest of Canada in Confederation—gold production had dwindled to $1.4 million.

By the early 1880s salmon had replaced gold as British Columbia's export staple, but not for long. The completion of the Canadian Pacific Railway in 1886 coincided with the discoveries of coal and base metals—lead, silver, zinc—in the Kootenay area. The mining industry boomed, as did the forestry industry based on British Columbia's immense stands of Douglas fir, spruce, cedar, and hemlock. By 1911 the fishing industry ranked a low third.

These three extractive industries required massive injections of foreign capital. Underground mining and ore smelting, timber felling and sawing, salmon cannery production, all required large, costly machinery—and a reliable labor force. The Indians had worked well in the fur trade, but by temperament they were unsuitable for industrial work. Theirs was the rhythm of the seasons: a time to hunt and a time to fish.

The industrial work force came from elsewhere: skilled workers from Great Britain and the United States, unskilled laborers from Asia and eastern Europe. They were an awesome army pitted, as they themselves saw it, "against the wilderness." In their words they were "taming the wild frontier."

The immigrants of a hundred years ago did not see themselves as despoilers and land rapists; they were simply miners, loggers, and railway builders. It was a golden age of capitalism, of unfettered free enterprise. Profit was sacred. Did nature not exist but to serve man?

Canadians like to think the assault was less ferocious north of the United States border, that because of some mythical virtue of restraint and forbearance they took better care of their rivers, mountains, forests, and fish. They delude themselves. It is true that the mistakes—the ecological disasters—occurred first in the States, but only because there were more people living there.

The fate of the salmon illustrates this point.

Prior to the white man, shimmering hordes of salmon thrashed their way upstream to spawn in almost every coastal river and creek from California to Alaska. In 1864 the first commercial salmon cannery on the West Coast began operations on the Sacramento River. Cases of canned Pacific salmon were shipped to England and elsewhere for handsome profits. Unsuccessful gold miners, or successful ones with stakes to invest, rushed into the fishery. As both fishermen and canneries multiplied, the salmon declined.

The great Sacramento runs were decimated by new fishing methods. Industrial technology had been brought to fishing. Fish traps—huge

V-shaped enclosures of netting mounted on piles extending far out in the estuary—could capture an entire run of salmon. Fishing wheels, which mechanically dipped the salmon out of the river, operated 24 hours a day. There were drift nets, seine nets, traps and weirs, gang hooks and barricades.

Miraculously, some of the salmon escaped to continue up the Sacramento to their spawning grounds. But as technological man learned to look for gold by hydraulic mining, thousands of tons of earth were flushed down the mountainsides to smother in mud, or to scour away, the delicate gravel of the spawning beds. When the salmon arrived, there was no place for them to spawn.

In a few years the salmon runs of the Sacramento River were gone. The canneries, and the fishermen, moved northward to the next big salmon river. One by one the salmon runs were destroyed by overfishing and through loss of habitat. The catastrophe was due to the greed of fishermen and canners, the heedlessness of miners and loggers.

The same destruction was taking place in British Columbia, where the primitive logging industry was extending its operations through nearly every watershed. Creeks were choked with brush from limbed trees. Spawning beds were blocked by logs or looted of gravel to make logging roads. Overcutting on hillsides removed so many trees that water runoff patterns changed. Smaller streams became mere drains. In summer they might be dry; in winter they could become raging torrents that ripped out gravel beds.

By gouging, log drives altered river beds, making them useless as spawning sites for salmon. Still other gravel beds were smothered in silt from eroding banks and hillsides denuded of protective ground cover. The washings of rain across gravel logging roads also carried silt into spawning beds.

Worse, the loggers blockaded the salmon streams with "splash" dams. These crude dams stored water which could be released suddenly, in order to "flash float" logs downriver. Walls of white water and tumbling logs tore up the gravel beds, destroying the salmon eggs there. In between these man-made floods, streambeds were nearly dry; in winter the eggs were often frozen.

Dam building was an easily recognized threat to salmon survival, yet it was permitted to continue. It was a time of the independent, "rational" man, the free enterpriser. Interests—in this case mining, logging, and fishing—were allowed to conflict. The excuse was "jobs and money." People and payrolls over trees and fish.

In 1898 a mining company built a dam across the Quesnel River at the outlet of Quesnel Lake. The intention was to impound the lake's entire outflow during late summer and early fall to expose the river bed for gold mining. As a consequence, salmon migrating to the spawning grounds of the Quesnel, Horsefly, and Mitchell rivers died in the thousands on the dry stones of the streambed. The runs to the Quesnel watershed have never fully recovered. A quarter of the productive rearing areas of the Fraser River salmon fishery was sacrificed to a mining venture that was itself a failure.

A smaller dam brought equal havoc to the famed Adams River sockeye run. In 1904 a logging company built a rock-fill, timber-crib dam at the top end of the Lower Adams River, at the outlet of Adams Lake. The splash dam was used to flash float logs downriver, and it left the river bed dry most of the time. Salmon eggs that did not succumb to frost and drought were torn up in the log-tumbling flash floods.

Such events are ecological disasters. Whole races of salmon were obliterated or depleted to the point of no recovery by industrial man's activities. They were disasters because the salmon must return to the freshwater streams of their origins to spawn. And this is in man's territory.

The loggers of a hundred years ago may not have known about the salmon eggs in the gravel. It is harder to find excuses for the cannery operators whose wasteful practices shocked people's sensibilities even then.

At the time, red-meat sockeye was preferred. Since fishermen's nets did not discriminate between the different species of salmon, thousands of dead chinook, chum, and pink were flung through the offal holes in the cannery floors into the muddy Fraser River below. In the 1870s most parts of the sockeye were thrown out with the offal too; it was easier to slice fillets from the sides and discard the rest.

The fishermen brought in boatloads more than the canneries could handle. Desperate to sell, the fishermen offered their catches for as little as a quarter of a cent a pound. When the canners refused them, the silver harvest of sockeye was summarily dumped by the dockside.

These dead fish, plus the thousands of tons of guts, heads, and tails, not only caused an unbearable stench for miles around but also created a health hazard. A typhoid fever outbreak in 1877 was attributed to the unsanitary conditions caused by the wastes from salmon canneries along the Fraser.

The incoming tides and the prevailing westerly winds kept the mess in the river, even moving it upstream where it washed ashore. The banks of the Fraser were lined, it is said, with a two-foot deep layer of silver slime—rotting fish and offal—which sometimes extended in a band 6 m (20 ft) wide. Farmers for 30 km (18 mi) up the Fraser pitchforked the dead salmon onto their fields and plowed them in as fertilizer.

The assault on the great salmon runs began in the early 1860s. By 1870

Sockeye salmon *(Oncorhynchus nerka)*

Historically sockeye is our most valued fish because of its red flesh which is rich in oil and retains its color and flavor under all conditions of storage. Sockeye spawn in streams that have lakes in their watershed, the young spending from one to three years in a lake before migrating to sea. In their third year at sea they begin sexual development and between this time and the sixth year, they return to their natal spawning stream. Mature four-year-old sockeye average 3 kg; older sockeye often reach 5.5 kg.

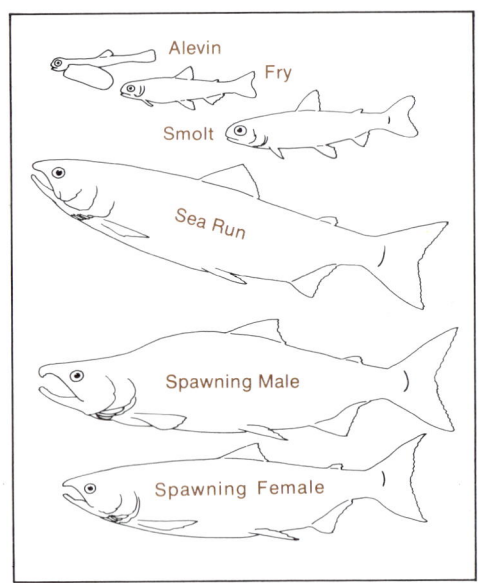

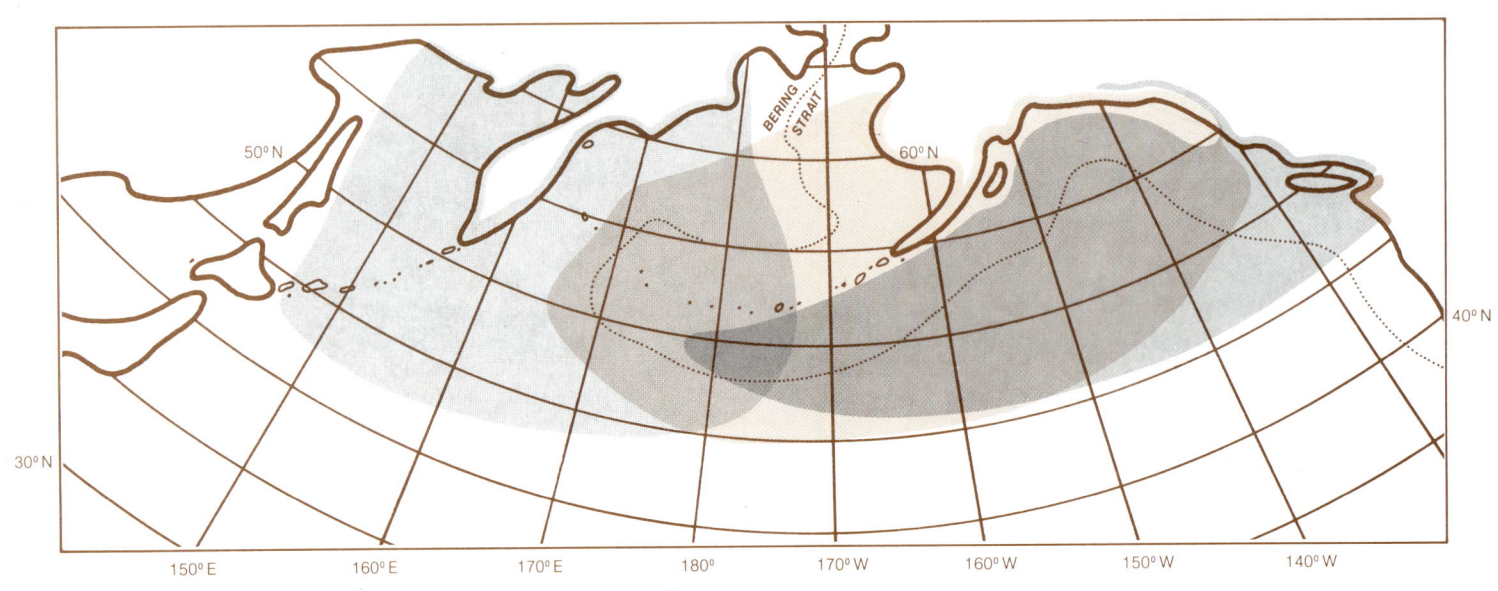

Pacific Ocean distribution

.... 200 Mile Limit

- **ASIA**
- **ALASKA**
- **BRITISH COLUMBIA**
- **WASHINGTON & OREGON**

Pink salmon *(Oncorhynchus gorbuscha)*

The smallest of the Pacific salmon, the pink lives only two years and has the simplest and least varied life history. Pink salmon always spawn as two-year-olds in both large and small river systems. When the young hatch and emerge from the gravel the following spring, they at once drift downstream to the sea. Their year and a half of ocean growth is rapid. Mature pink average 2 kg in weight; a few 4.5 kg pink have been caught.

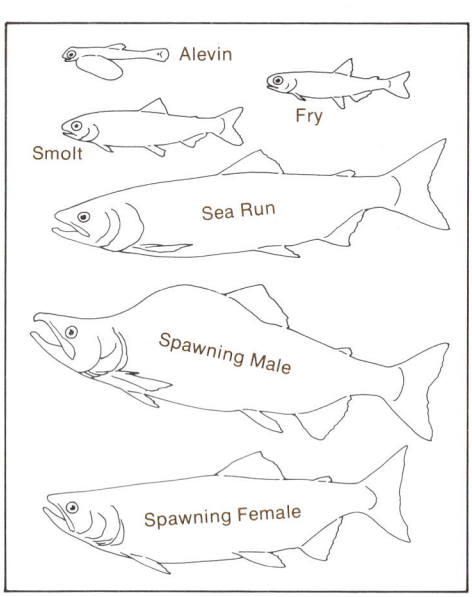

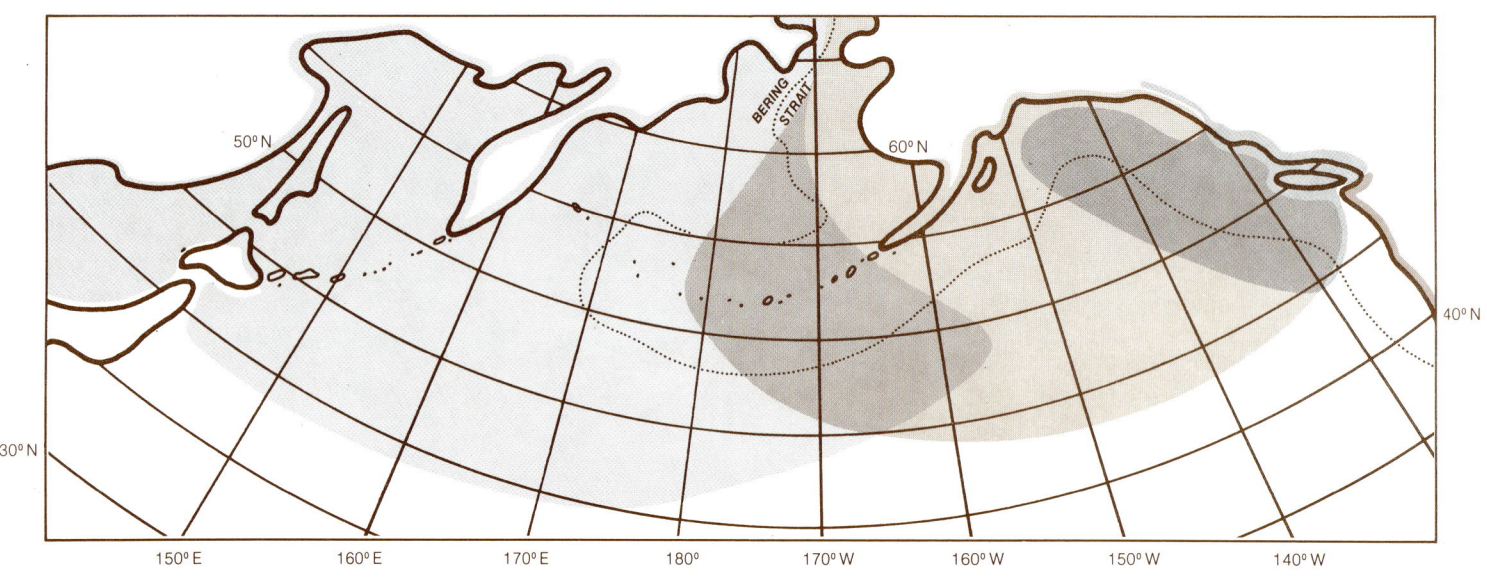

Pacific Ocean distribution

.... 200 Mile Limit

- ASIA
- ALASKA
- BRITISH COLUMBIA
- WASHINGTON & OREGON

Coho salmon *(Oncorhynchus kisutch)*

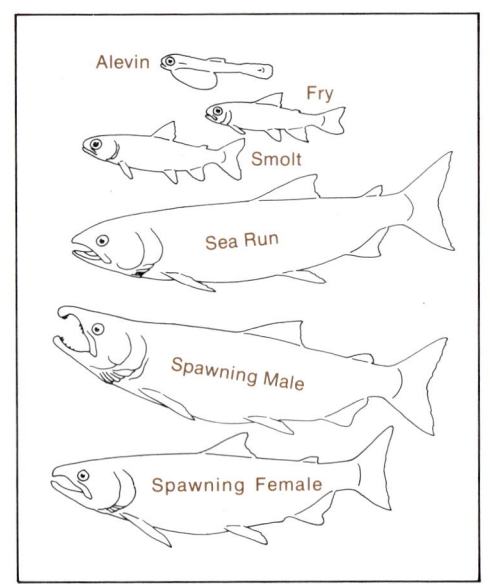

Popular as a game fish, coho are fast, strong and magnificent jumpers. Like the pink and chum, the coho rarely go far inland to spawn, preferring coastal creeks close to the ocean. Next to the pink, the coho probably has the simplest life history: one year in the nursery stream, two years in salt water. A few coho spend three years in the ocean, but these are the exceptions. By the time they return to spawn, they weigh 4.5 to 6.5 kg.

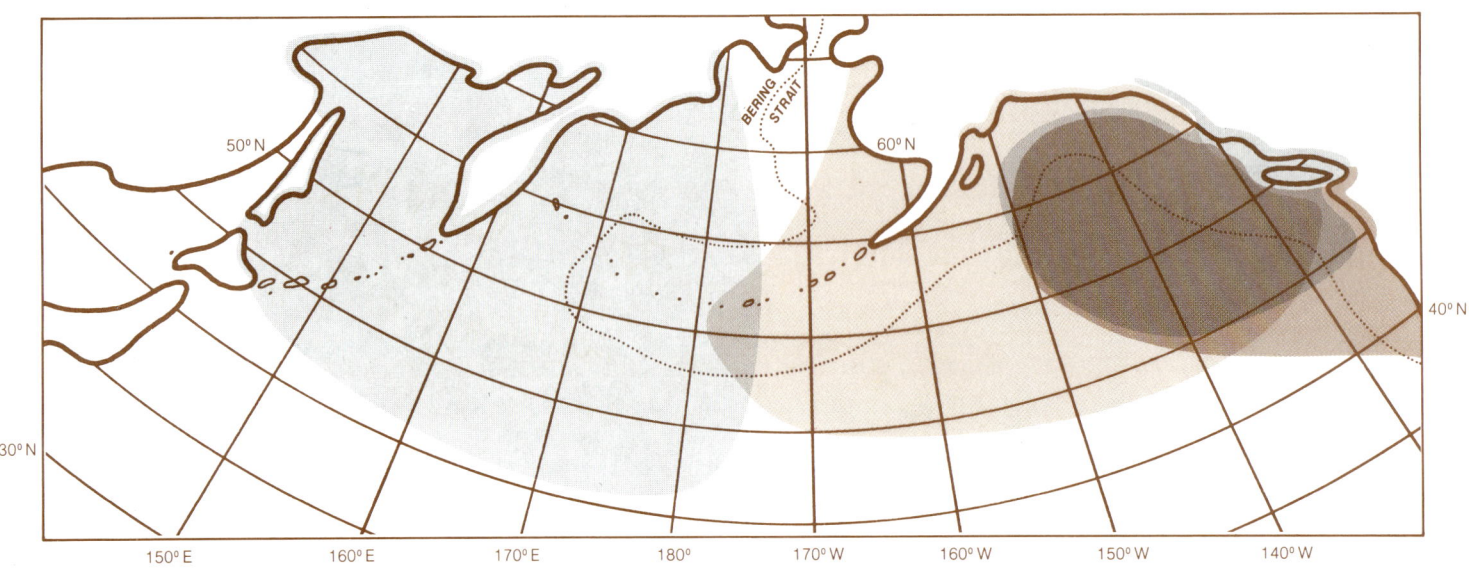

Pacific Ocean distribution

.... 200 Mile Limit

■ ASIA
■ ALASKA
■ BRITISH COLUMBIA
■ WASHINGTON & OREGON

Chinook salmon

(Oncorhynchus tschawytscha)

The largest of the salmon, chinook commonly average between 14 and 18 kg in weight, and record fish weighing 56 kg have been caught. They also live the longest and have a varied life history. Chinook reach maturity at between three and seven years. Most are four to five years old when they return to spawn. Although most favor the spring or fall migration, they can be found heading for their upstream spawning grounds almost any month of the year. They prefer major rivers such as the Columbia and the Yukon for spawning and their grounds can extend from a few kilometers inland to 1600 km or more. Most young chinook fry go to sea soon after hatching, but others remain one or two years in lake or river before making their way downstream to salt water.

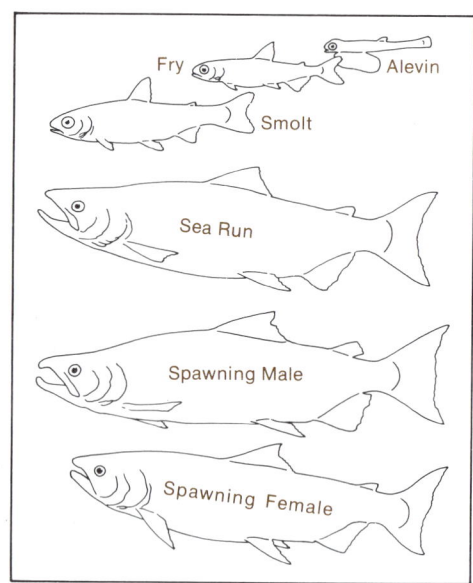

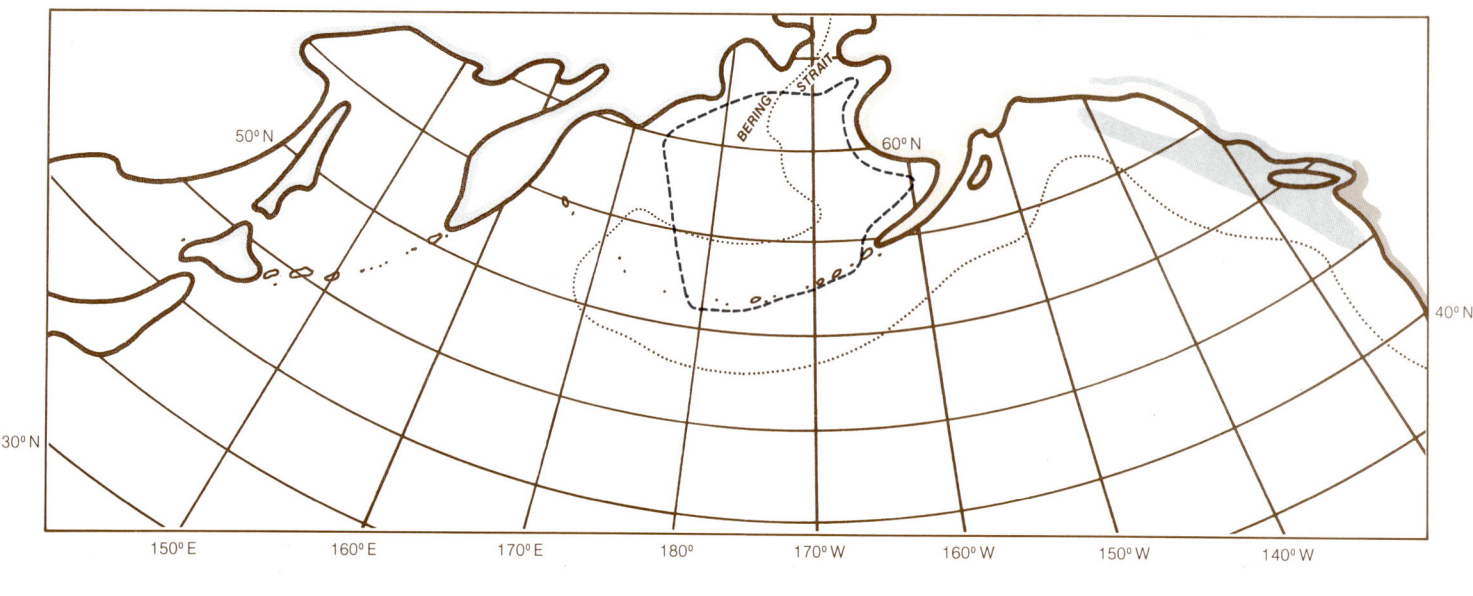

Pacific Ocean distribution

Chum salmon *(Oncorhynchus keta)*

The last of the Pacific salmon to enter fresh water, chum arrive in coastal streams in the late fall although there are summer chum runs to northern British Columbia streams as early as July. Spawning fish of this species do not usually travel far inland to spawn, the exception being Yukon River chum, which travel 3000 km upstream before stopping to spawn. Like the young of the pink salmon, chum start for the sea almost as soon as they emerge from the gravel as fry. The chum reaches maturity in the third or fourth year at a size ranging between 3.5 and 4.5 kg.

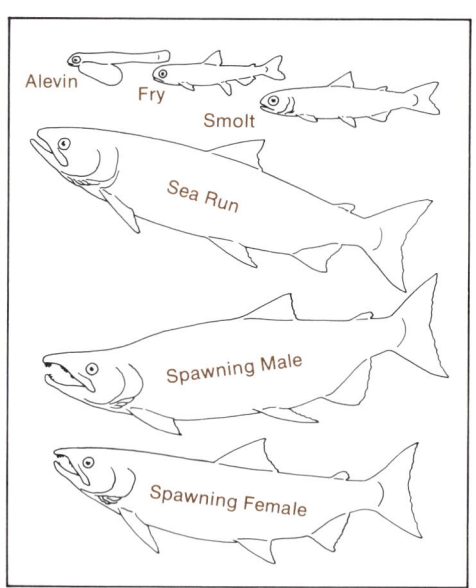

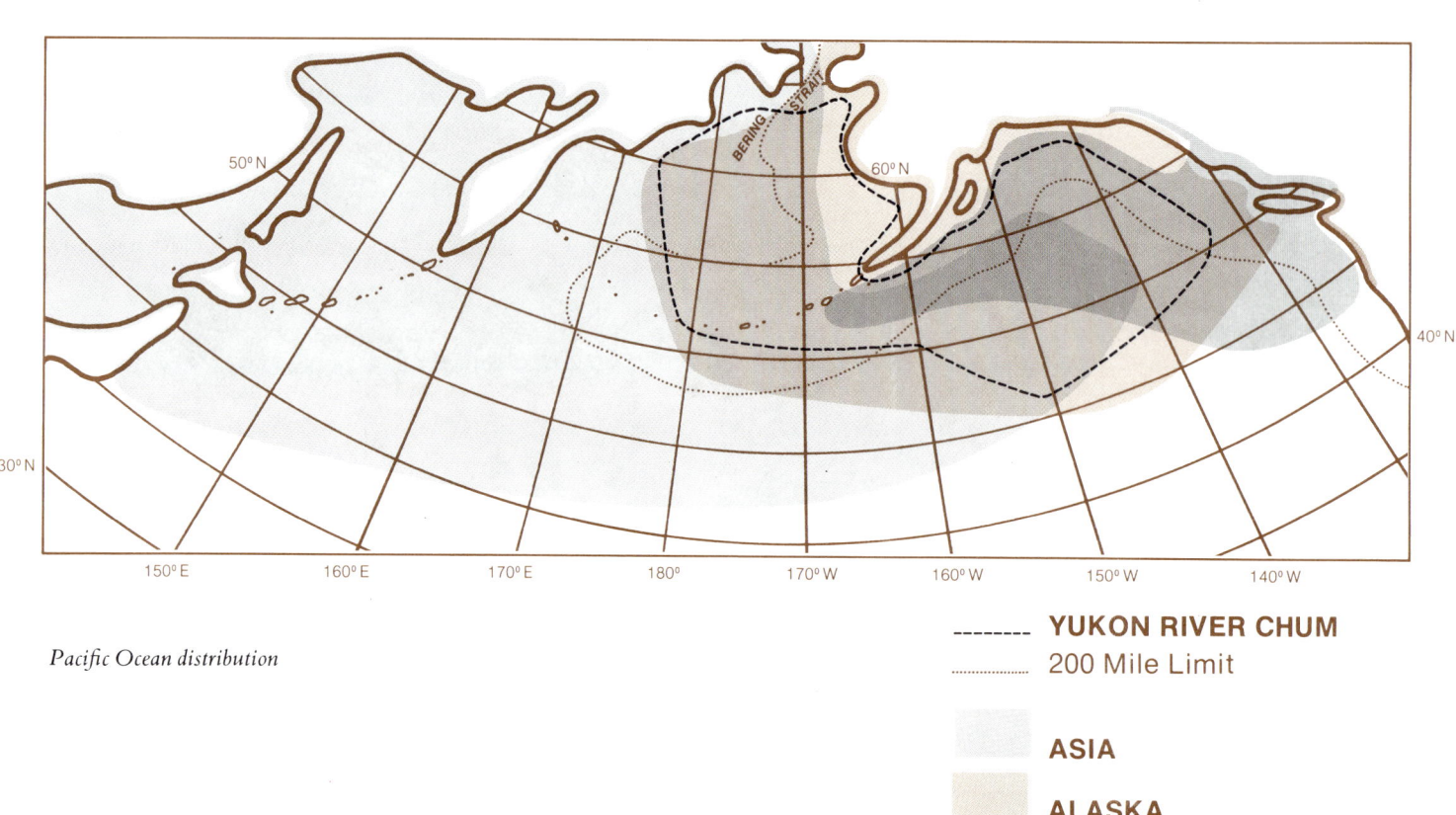

Pacific Ocean distribution

-------- **YUKON RIVER CHUM**
.... 200 Mile Limit

ASIA

ALASKA

BRITISH COLUMBIA

WASHINGTON & OREGON

Steelhead trout *(Salmo gairdneri)*

More closely related to the Atlantic salmon than to the five Pacific species, the steelhead, like their Atlantic relatives, do not always die immediately after spawning; some survive to return to the sea. Individual steelhead have been known to spawn three times. The fry emerge from the gravel to live in the spawning stream for one or two years before migrating to the ocean. They return as adult spawners in their third, fourth or fifth year of life. Most of them will be caught by sports fishermen in their home stream. The largest steelhead can reach 115 cm in length and weigh as much as 17 kg.

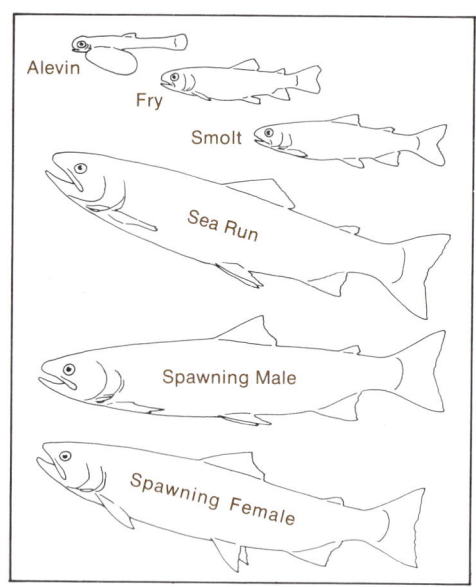

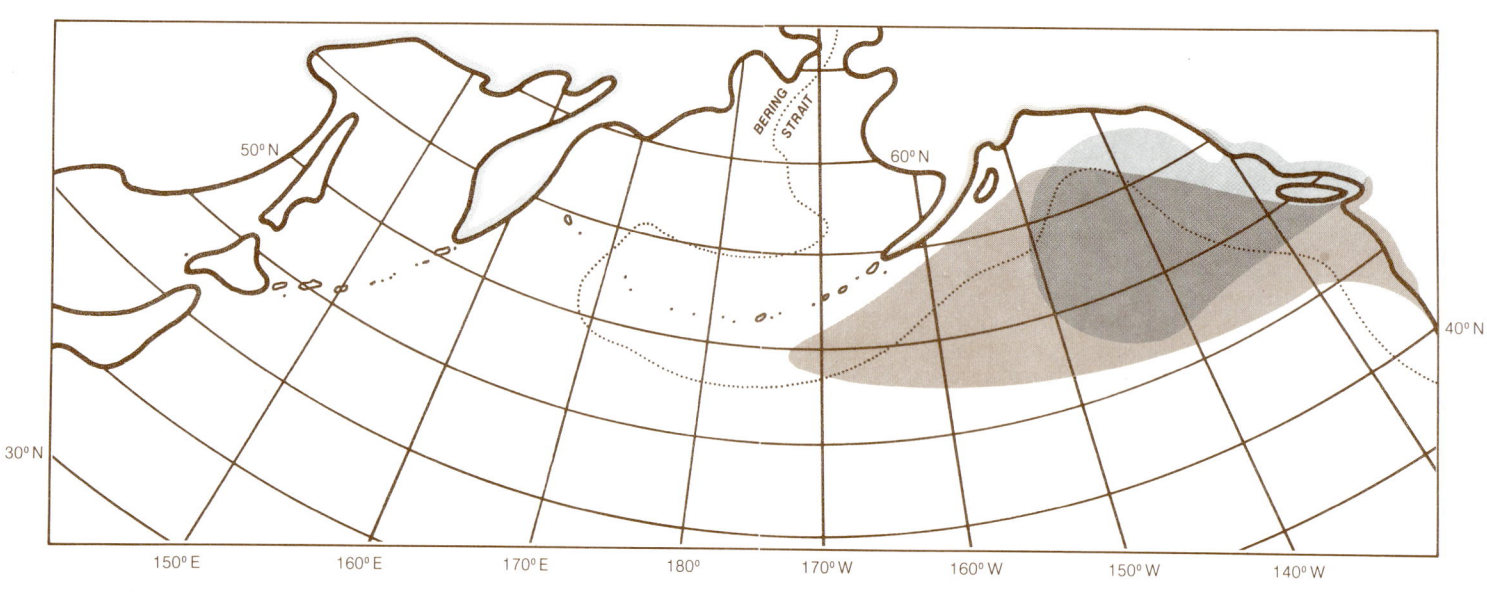

Pacific Ocean distribution

·········· 200 Mile Limit

- ASIA
- ALASKA
- BRITISH COLUMBIA
- WASHINGTON & OREGON

the results of overfishing were apparent; the salmon runs were in sharp decline. It took *two decades* for the government to react. In 1889 the Canadian government enacted legislation, in part to control abuses of the salmon industry. Under "Section 1—Salmon," the regulations read:

1. Fishing by means of nets or other apparatus without leases or licences from the Minister of Marine and Fisheries . . . is prohibited in all waters of the province of British Columbia. Provided always, that Indians shall at all times have liberty to fish for the purpose of providing food for themselves, but not for sale, barter or traffic, by any means other than with drift nets or spearing.

3. (a) Drifting with salmon nets shall be confined to tidal waters, and no salmon net of any kind shall be used for salmon in fresh waters.
 (b) Drift nets shall not be used so as to obstruct more than one-third of any river.

The General Fishery Regulations for British Columbia came into effect on 18 July 1889. The first offense recorded was that of a fisherman in Burrard Inlet fishing with dynamite. The astonished fisheries officer was told that it was the standard method of fishing in the inlet. The explosions stunned the salmon and they floated to the surface where they were easily picked up. (This is not an efficient way to fish; most of those killed by an underwater explosion will sink to the bottom.)

The most destructive of the white man's new methods was the fish trap, a permanent installation of piles and netting, extending a half-mile or more across the line of a homeward migrating salmon's travel. As with all such herding traps, it led the fish into increasingly smaller enclosures until, finally, there was no way out and the salmon could be easily taken. During large runs these traps were known to fill faster than the owners could unload them. On one such occasion in Puget Sound, tens of thousands of fish were wedged so tightly in a trap that they suffocated, destroying one of the sockeye runs. At the time, no one noticed; there was always, it seemed then, another and larger run of salmon to come.

It was an era—that last third of the nineteenth century—of ruination of the salmon. The industry was to blame. Overfishing and willful waste were normal practices. The Americans ruined *their* great runs: the Sacramento, the Rogue, and the mighty Columbia. They then turned their nets to catching Canadian-bred salmon inbound through United States waters on the way to their Fraser River spawning beds. This is an irksome issue today, but in the late 1800s few thought the Fraser River salmon runs could be hurt by overfishing.

The reason for this optimism was that people knew that one pair of adult spawners could reproduce several thousand young. It did not take many adults escaping the fishing nets and completing their journey to the spawning grounds on the upper reaches of the river to replace each year's silvery salmon bonanzas in the straits, estuaries, and river mouths.

There were informed people then, as now, who were conservation-minded. These people allied themselves with fisheries officers in efforts to remove man-made obstacles to migration. Stream clearance programs were undertaken. Some waged war on predators such as bears and squawfish. Others rescued stranded salmon smolts from riverside pools, putting them back in the main stream so that they might continue to the sea. Later in the year some people might be found netting and carrying adult spawners past obstructing waterfalls.

Their efforts, given the enormous number of fish in the overall river system, were of marginal value, but they made a lasting contribution by impressing on the public the value of helping the salmon.

Biologists of a century ago knew much about the salmon, their knowledge being based on studies of the Atlantic salmon (genus *Salmo*), a close relative of the Pacific salmon (genus *Oncorhynchus*). There are six species belonging to *Oncorhynchus*, only five of which are found in British Columbia waters; their common names are: sockeye; pink; coho; chum; chinook. The Japanese cherry, or masu, salmon is limited to the Japanese islands and the nearby Asian mainland.

Their scientific names are Russian in origin. The species of Pacific salmon were first identified on the Asian side of the ocean by Georg Wilhelm Stellar, a German surgeon-naturalist employed as a teacher at the Saint Petersburg Academy of Sciences. In 1737 the young German joined Vitus Bering's second expedition to Siberia, which had departed four years earlier. Traveling alone and mostly on foot, Stellar crossed the Urals, the steppes and Siberia, taking three years to complete the 8000-km (5,000-mile) trek. He caught up to Bering on the east coast of Kamchatka Peninsula in time to accompany the Danish sea captain across the mist-shrouded Bering Sea to Alaska. They were the first whites to land there.

On the return voyage they were storm-wrecked off Bering Island, a narrow, treeless strip of sand almost within sight of Kamchatka. Their plight was made more desperate in December 1741 when scurvy took the lives of many of the crew, including that of their discouraged captain. Thanks to the ingenuity and leadership of Stellar, the starving survivors made it through that dreadful winter. In the late spring, salmon returned to the rivers, and the expedition members regained their strength. A new packet was built which got them the remaining distance to Kamchatka.

For the next three years (1742-44) Stellar explored Kamchatka and northeastern Siberia, usually by himself and under conditions of extreme hardship: it was a feat of scientific exploration. He not only collected innumerable plant and animal specimens, but also wrote voluminous notes and draft papers on natural history. Among other things, he observed and identified the different species of Pacific salmon. In 1744 Stellar died of his exertions at age 37, on his way back to Saint Petersburg.

Another scientist attached to the Bering expedition, Stefan P. Kracheninnikov, wrote a natural history of Kamchatka ten years later. Much of this book was based on Stellar's notes and journals, including his work on the salmon. Ultimately, Johann Julius Walbaum, a German ichthyologist, used Stellar's vernacular Russian names for the salmon in his *Artedi Piscium*, published in 1792.

In the proper form the name of the genus—*Oncorhynchus*, or its abbreviation *O.*–should stand in parentheses before each of the species names; thus: sockeye *(O. nerka)*; pink *(O. gorbuscha)*; coho *(O. kisutch)*; chum *(O. keta)*; chinook *(O. tschawytscha)*; and Japanese cherry *(O. masou)*.

Kracheninnikov's book, published in 1755, was translated into English in 1764. Thus, by the time the commercial salmon fishery was established in British Columbia in the 1870s, a great deal of the life history of *Oncorhynchus* was already known. It was common knowledge that the salmon are born in fresh water but live most of their lives in the ocean, and that they return to the freshwater streams to spawn and afterward die. It was known that salmon eggs are deposited in the gravel of streams and lakes by the female and are fertilized at the same instant by the male. Further, it was known that each spring the eggs hatch and the fry emerge from the gravel beds.

Some details were less understood; for instance, that whereas all pink and most chum fry migrate directly to salt water, those of coho, masu, and sockeye may stay up to three years in lakes or streams before descending to the sea. Chinook fry generally leave fresh water before their first winter.

Where the young salmon went once they reached the ocean was a mystery. Early investigators speculated that the salmon stayed close to shore near the mouth of their home river. On reaching maturity they could then easily return upriver to their ancestral spawning grounds.

The importance of the spawning grounds was recognized by the authorities responsible for the British Columbia fishery. A systematic mapping of rivers and spawning grounds was underway by the early 1870s, when fisheries inspectors were keeping records of production in some of the major spawning areas.

It was realized by these early observers that most salmon mortality occurred during the earliest stages of life; that is, during the incubation of the eggs and development of the fry. They concluded that protection of these early stages would result in many more adult fish. The first successful fish incubation experiments in Canada were conducted with Atlantic salmon on the Miramichi and Restigouche rivers in New Brunswick in 1874.

Excerpts from the Annual Reports, Department of Marine and Fisheries indicate similar activity in the West:

> The question of a salmon hatchery on the waters of the Fraser continues to be agitated. By telegram from the Department, of 10th July last, I was asked to "name the place on the Fraser River best suitable for a salmon hatchery?" To this I telegraphed in reply, suggesting that, in a decision so momentous, the opinion of an expert from Canada, cognizant of all the requirements, should rather be called for.
> (1882)

> As instructed, I engaged, in October last, Mr. Thomas Mowat, to choose a site on the Lower Fraser for the erection of a fish hatchery. The site has been duly selected at a point some four miles above New Westminster, on the opposite shore, where Messrs. B. Haigh & Sons have a cannery.
> (1883)

> A fish hatchery, measuring 100 x 40 feet, was built during the last season on the Fraser River. It will easily accommodate 3,000,000 *Quinnat* salmon eggs, or 5,500,000 *Saw-quai* salmon ova. By doubling the trays, double this number of eggs can be laid down. The catching of parent salmon began about the beginning of July, and by the close of the season 3,000,000 eggs were on the trays. The operations, though novel, were highly successful and reflect credit upon the officer in charge, Mr. Thomas Mowat.
> (1884)

This first Canadian hatchery for Pacific salmon was built in 1883 at Bon Accord on the Fraser River, near the present site of New Westminster and was followed by others at Granite Creek, 1902; Lakelse Lake, 1903; Harrison Lake, Rivers Inlet, and Pemberton, 1906; Babine and Stuart lakes, 1908.

Most of the effort in early British Columbia hatcheries was devoted to increasing the available numbers of sockeye salmon. This was the most valuable species, and obvious declines in its catch were occurring before the end of the century. The objective during this period of about 20 years was to take and incubate as many eggs as possible, and to release the resulting millions of fry to fend for themselves. Along the West Coast, hatcheries each year turned out many millions of salmon fry and fingerlings that were planted in their streams of origin. As well, fish were transplanted to other watersheds so that new populations could be established, or so that off-year runs could be filled in.

In the late 1800s millions of chinook, sockeye, pink, and coho salmon eggs were planted in streams in Europe, Hawaii, Australia, New Zealand, Argentina, Chile, Mexico, and Nicaragua. Even greater transfers were made from western North America to the streams of eastern Canada and the United States.

Of all the transplantings of Pacific salmon made in these early years, only the chinook and sockeye taken to New Zealand's South Island can be said to have been successful, although the seagoing sockeye run changed to the landlocked kokanee. Chinook and sockeye thus became the first Pacific salmon to establish themselves in the southern hemisphere.

Although salmon were planted in a great variety of conditions, the freshwater and marine habitats in the new territories seldom matched those of their native range. The monumental efforts of the 1870s through early 1900s to transplant Pacific salmon were all but futile, yet the promise remained; the work went on. By 1929 searun populations of chinook had been established in the St. John River of New Brunswick and the Port Credit River of Ontario. In 1938 a U.S. scientific journal could report: "The streams and coastal regions of Maine, New Brunswick and Ontario are the only foreign waters on the North American continent in which natural populations of the Pacific salmon have been established."

These runs soon petered out.

Plantings of Alaskan pink salmon in Maine seemed to be successful, but later the runs dwindled away. Rivers of northern Europe that seemed ideal for the Pacific chinook saw none return from the seaward migration of smolts. These same rivers supported large stocks of native trout and Atlantic salmon.

Non-anadromous rainbow trout soon became the most universal salmonid. The hardiness of their eggs and their passivity in confinement made them easy to transport, and their adaptability to temperate fresh waters of either the northern or southern hemispheres made them excellent colonizers. They are now established in wild and cultured form on every continent except Antarctica. Their only salmonid rivals for universality are the European brown trout and, more recently, the kokanee.

Later salmon transplants were more deliberate. Biologists attempted to find donor streams whose freshwater conditions matched those of the receiving areas. More transplants were made within the salmonid's native range. Sockeye eggs were transferred from Afognak Island, Alaska, to the Fraser River. Within the Fraser itself, desperate attempts were made to replace sockeye runs that had been destroyed by the Hell's Gate rock slide in 1913, by transferring eggs, fry, and smolts from one tributary to another within the watershed, and from outside areas such as Babine Lake on the Skeena River system.

Transfers of chinook, coho, and sockeye salmon between hatcheries in Washington, Oregon, and California were so numerous—and so poorly documented—that it is not possible today to compile all the information. In some cases hatcheries were able to maintain runs of transplanted salmon where none or few had been before, but there are not many places where it can be proven that a new wild stock has become established as a result of transplanting.

The efforts of biologists to combat the declines in the salmon resource by propagation in hatcheries and transplants of eggs had been underway for half a century with little effect. One problem was that "success" was virtually unmeasurable. There were wide variations in the numbers of salmon from year to year and from generation to generation, and much of this variation stemmed from weather fluctuations or from complex biological causes that are even now not well understood. Moreover, the effects of the commercial and sports fisheries for salmon were largely unknown, being only partially revealed by the annual gross statistics of catches.

During the period 1875 to 1925 hatcheries proliferated in North America, and British Columbia was no exception. With experience came the realization that if the fry were protected for a little longer, or fed and released at a larger size, perhaps more would survive. Starting about 1906 small earthen ponds were excavated to provide a place to rear the fry. Early diets consisted of ground fresh fish, salt fish, beef liver, or hearts. In some instances supplementary feeding in ponds was achieved by suspending fish carcasses or cattle heads on racks over the ponds so that the maggots would drop into the ponds to be consumed by the fish.

Attempts at rearing fish led to other problems, primarily fish disease. When fish are confined in higher than normal densities, their susceptibility to infectious diseases increases. Hatchery practice became one of rearing fish to as large a size as possible before too many died. When the mortality rate rose dramatically, the fish would be released before any more died on site. The fish so produced were of poor quality, and most failed to survive.

The practice of large egg takes and hatchery rearing continued until the 1930s. Then, investigations by biologists of the Fisheries Research Board of Canada showed that the annual release of millions of hatchery-bred fry was making no detectable contribution to the catch of sockeye salmon, nor was there any noticeable increase in the numbers of spawners returning to their home streams. By the late 1930s sockeye hatcheries had fallen into disrepute. All the salmon hatcheries in British Columbia were phased out, and many American facilities were either closed or left operating under drastically reduced funding.

Although salmon hatcheries in British Columbia had been phased out in the Depression years because of fish disease and lack of funding, there remained in Washington, Oregon, and California a small nucleus of fish culturists who firmly believed that the basic premise was still correct: by eliminating the high mortality rate that occurs in eggs and fry in the natural state, it should be possible to produce large numbers of adults. They felt that if they reared the fish for a longer period and to a larger size before release, the subsequent survival would be much higher. Two developments during the war years of 1939 to 1945 improved the prospect of salmon rearing: nutritional research resulted in much improved hatchery diets, and the development of inexpensive antibiotics made their use practical in the treatment of some fish diseases.

By the early 1950s chinook and coho were being raised in U.S. hatcheries to a larger size and with fewer losses to the common diseases. It was now possible to release healthy fish of a size approaching that of wild smolts. Further refinement of diets began to result in increased production of adult salmon. By the late 1960s, the dream of the early salmon culturists had begun to be realized: the hatcheries were producing spectacular results with chinook and coho.

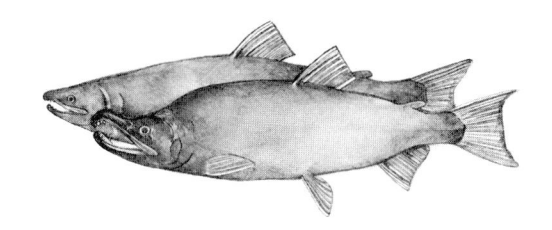

Migrating adults

Biology

Courting sockeye

Fighting sockeye males

Pink nest-building

Until recently almost all of man's contact with salmon has been with the final stages of its life, when the adult salmon return to fresh water to spawn, then die. The basic facts of salmon spawning and egg development were well known in the earliest days of the Pacific coast fishery. At the time, the resource seemed limitless, but by the end of the 1930s the trend was clear: the great runs were in decline. It was then that Canadian Fisheries scientists began a long series of studies of the five species of Pacific salmon. From these studies would come knowledge that might one day save the salmon from extinction.

Each of the salmonid species has its own life history, habits, and relative abundance. Each species divides itself into separate runs in different river systems, and yet again into diverse stocks or races within the watersheds. The salmonids vary in one particular aspect from most other fish: although much of their life cycle is spent in salt water, their dependence on fresh water is almost absolute. They need the freshwater rivers, the tributaries, the lakes. They need them not only for breeding but also for the incubation of the eggs and survival of their young. These must remain in the nursery streams and lakes until they migrate to the sea.

Years later, as adults, they leave the safety of the ocean to return, congregating in the shallow inshore waters near the mouth of the home river. When it is time, each stock begins an upstream odyssey that may take weeks. For some stocks of salmon the hereditary waters lie hundreds of kilometers from the sea. Theirs is a remarkable feat of endurance.

It is also a remarkable feat of homing. The ability of each salmon to navigate accurately the expanse of ocean, to return to its estuary of origin and finally "home" to a particular spawning ground far inland, is astonishing and as yet not fully understood. What is known is that each race and run of salmon is made up of uniquely specialized individuals. Each sexually maturing salmon will seek out the exact conditions—of temperature, gravel, and waterflow—that will assure the success of its offspring. To do this, the adult spawners are biologically "programmed" to return to the same tributary (or lake) and gravel shallow in which their parents spawned and they themselves were hatched.

The freshwater journey begins when the fish enter the river from the estuary. From here to their upstream destination they swim more or less as a group or school. They usually choose the deeper parts of the river. If the currents are swift, the salmon may follow along the banks in single file. The rocks, boulders, and shoreline contours break up the fast-flowing water into eddies and flows that the salmon can negotiate.

Whereas the salmon may swim 22 hours a day during the ocean part of the homing run, the river ascent is taken in easier stages. They pause in their migration to loiter in river pools. These periods of rest are often in the afternoons or during the hours of darkness. If conditions during the upstream climb are favorable, the salmon may reach their destination early.

The fish may school again, sometimes for days, holding in a dense, tight mass near the mouth of the spawning stream. It is a period of waiting for final maturation. The eggs ripen and separate within the females; sperm, called milt, forms in the males. When the spawners are ready they leave the undulating pack of milling fish and enter the shallower waters of the creek.

Females arriving on the spawning ground swim slowly along the bottom, touching the gravel with their extended lower fins. Males cruise the shallows, watching for unattended, nest-building females. They compete with other males by threat displays and occasional fights, brief flurries of splash and spray, teeth and tails. One female may be the preoccupation of three or four full-sized males and a bevy of small jacks (underage but sexually precocious male salmon). These are kept at a distance by the older, stronger males but often succeed in fertilizing some of the eggs.

It is not clear to what, exactly, a female responds in choosing a suitable site to build a "redd," or series of nests. Generally, her nest building will be accompanied by a courtship display by the male. She is the dominant partner; the male responds to her actions. The mating sequence, which goes on for many hours, starts with the selection of the nest area.

Many nests are built in riffles or just in front of logs and boulders. These obstacles cause increased waterflows, and here females are seen to suddenly stop their downstream drifting. They abruptly turn and swim upstream to test the gravel, first with their noses and sometimes with a few careful digs. Yawning may accompany these probes. The yawns signal a change of mood: the female now becomes aggressive, attacking other salmon that pass too closely to the territory she is surveying.

When a location is suitable, the female starts the second phase, that of digging a general outline of the nest. This is the signal for her consort to begin his courtship display. His position is beside the female, with his head

Courting chum

Female chum digging nest

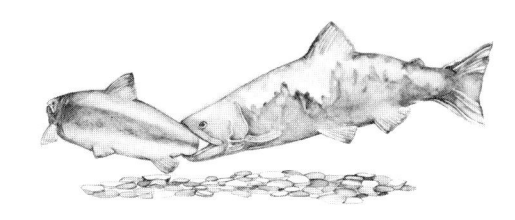

Male chum attacking uncooperative female during spawning

Male chum nosing female's anal vent, possibly checking her for spawning readiness

Female chum probing nest depth with anal fin

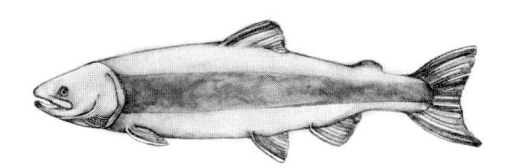

Female chum showing characteristic dark lateral band

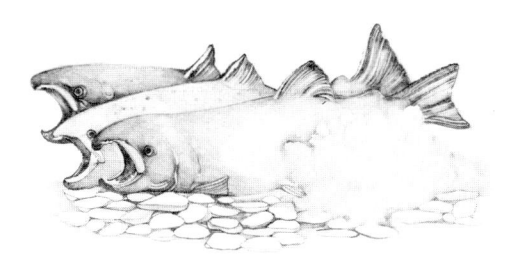

Two coho males fertilizing eggs from female spawning between them

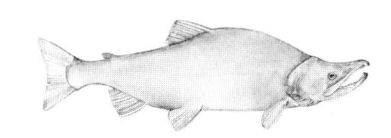

Sockeye jack (40.2 cm)

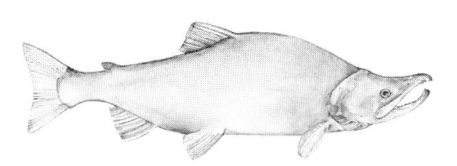

Mature sockeye male (58-66 cm)

at the level of her dorsal fin. The male trembles noticeably as he moves forward and alongside the female. He frequently switches position from one side of his mate to the other by crossing over her back. These characteristic movements are known as "quivering" and "crossing over."

From time to time the male interrupts his courtship to make sudden forays against encroaching adult males and the circling jacks. The jacks flee at once, only to return, but mature males often stay to challenge for the right to mate.

Less frequently but with more viciousness, the female darts from her chosen area to bite the tail of an intruder, be it female, adult male, or jack. Having driven off the unwelcome intruders she returns to circle the nest, attended by her solicitous mate. There is agitation and drama in the restless hovering circuit of the pair. Finally she returns to her work.

In digging, the female turns on her side and violently whisks the stream bottom with her tail. The forceful beating of the gravel, plus the partial vacuum created by her whisking, lifts stones off the bottom which are shifted downstream by the current. To avoid being propelled upstream by her own strong tail movements, she extends her pectoral fins as brakes against the current. Often her mouth is held open, another brake to forward impetus. (Males are seen to make digging motions when another male enters the nest, but these are of no help in the actual nest building. In his pseudo-digging, the male does not use fins and mouth as brakes, and so shoots ahead in the stream.)

This scene—the nervous circling of male and female, the ambitious challengers, and the anxious jacks—repeats itself. It takes hours, for the sweeping of the female's tail on the streambed moves only a few stones and bits of gravel at a time. After digging four or more nests, her tail fin will be badly frayed.

The construction of the nest is common to all the salmonids. The female first creates a semicircular depression in the streambed. As the depth increases, she gradually concentrates her digging at the center. This results in a nest pocket in which she will deposit her eggs.

Eventually, in the third phase of the nest building ritual, she tests the pocket depth by "probing." With her body arched in a U-shape, the female poises a few inches above the egg pocket. Then she drops downward until her anal fin touches or slips between the larger stones at the bottom. On contact she returns to her original position. Probing excites the male whose courtship behavior, especially quivering, increases.

SPAWNING

As oviposition—the laying of the eggs—approaches, events take place in predictable order. On completing a dig the female drifts backward so that her vent is over the egg pocket. There she begins a probe. The male pulls up beside her, trembling as he does so. The tension appears unbearable. Usually, the female completes the probe and lifts herself out of the nest. After crossing over her back one or more times, the male reverts to his original location, head level with her dorsal fin. This digging and probing goes on until the 16- to 18-inch depression appears to be sufficiently deep.

The phase four sequence culminates in spawning. In some species a dark lateral bar will develop in the pigment on the female's sides. This is the signal to her mate that she is ready to spawn. This time, instead of just probing the pocket, she remains there, openmouthed, allowing the male to assume the same position beside her. Mouths agape from muscular tension, fins vibrating, the partners release eggs and sperm together. The orange-red eggs settle downward in clouds of milky sperm. In seconds the seminal cloud dissolves and disappears in the current.

Sometimes this tableau is confused by the action of the attendant jacks who, at the very instant of spawning, rush into the nest to join the mating pair. Their combined donations of milt cloud the water a little longer and fertilize some of the settling eggs. When the water clears, the eggs are visible at the bottom of the nest. The female begins to cover them at once. This begins phase five.

Covering is a different type of digging. Here, the female moves directly upstream from the egg pocket, turns on her side, and, after laying her tail on the gravel, moves it upward in a fanning motion. The first few tail beats are gentle; they normally do not move any gravel. Instead, they create currents which lodge the eggs in the interstices of the larger rocks. Her digging gradually becomes more vigorous and blends into the excavation of the next nest; that is, she starts again at phase two.

In the shallow gravels of numberless streams in the late summer and fall, the same sequence of events is repeated. Each female digs a succession of nests, the covering of one being the start of the next. In each nest she will have laid from 500 to 1,000 eggs. These nests are guarded by the female until she dies some days later. During this guarding period—phase six—the female will continue digging, adding to the whole mound of gravel covering all her nests. She is belligerent toward other fish, especially other females attempting to dig in areas near her nest site. These she continues to attack until, weakened by approaching death, she drifts downstream.

All Pacific salmon deteriorate and die soon after spawning. The deterioration starts in the ocean with the onset of sexual maturity. All the glands appear to be affected, but especially the pituitary and adrenals. When the ritual of courtship and spawning is completed, there is a rapid aging of the salmon's body in all its parts: blood, tissue, and organs. Death results from

Social hierarchy of chum males indicated by body markings. Vertical pattern denotes dominance and horizontal pattern denotes subordinance

Dead salmon after spawning

Glaucous-winged gull

extreme acceleration of glandular activity. For sockeye salmon the average time between the start of spawning and death is nine to ten days.

The stream becomes filled with dead and dying salmon. They are seen gasping in the shallow, quiet places. Finally, no longer able to maintain their swimming position, they roll upside down. They may become wedged between shoreline rocks, or the end might come amid the trailing branches of a fallen tree on a stony beach or gravel ledge. In the still, out-of-the-way places they await death, eyes glazed and gill covers laboring for oxygen. Or they drift helplessly, unable to fight the current or even to remain upright.

The living unspawned salmon swim or hold position amid the shadowy cadavers, harbingers of their own demise, to which they pay no heed.

Sunny autumn—color in the leaves, salmon in the water, gulls skimming and squawking—is a beautiful picture. Later, in the drizzling winter rains, the stench of rotting fish pervades the woods and the scene is one of death and desolation. Windrows of flattened salmon lie eviscerated, walked upon by fat and quarrelsome gulls. The banks and sandbars for miles are littered with eyeless corpses, salmon eyes being choice morsels for the scavenging birds. Eagles, gulls, and crows feast on the flesh and entrails all winter long. It is hard to remember that this is nature's process, that the stink and decay is part of a cycle, that essential elements are being returned to the stream for life to come.

Part of that life is already there, safe in the gravel beneath the flowing water: the next generation of salmon. The miracle of the second generation began at the instant the parent salmon crouched together in the nest, jaws agape, bodies taut with strain, expelling eggs and sperm.

FERTILIZATION

Even as the reddish-orange eggs drift in a whitish cloud of milt to the bottom of the nest, fertilization is taking place. It is an astonishing phenomenon. While in the male, the ripe sperm cells are immobile, suspended in the semen. On exposure to water these inert, tadpole-shaped spermatozoa become frenziedly active, especially in the presence of coelomic fluid, which is the liquid that lubricates and separates the eggs while inside the female. The sperm cells' time for action is brief. During these critical seconds they must find and enter the funnel-shaped opening, or micropyle, of the egg. The egg assists them with a chemical guiding agent exuded from the micropyle.

Egg and sperm

Many spermatozoa, propelled by their madly lashing tails, may enter the micropyle, but only one can penetrate the thin membrane surrounding the inner egg. Here the swimming spermatozoon discards its whip-like tail. The head, or germ cell, travels inward to unite with the egg nucleus. The joining of the two nuclei is the actual fertilization.

Meanwhile, other remarkable processes are underway. The pearl-shaped egg is "water hardening": it is absorbing water and beginning to swell. This swelling closes the micropyle (preventing further access to spermatozoa), causes the egg to harden, and makes the surface of the egg seem sticky. This adhesive effect is due to suction of water through the capsule wall; it helps anchor the eggs to the stream bottom while the female covers her nest with a protective layer of gravel.

In an hour the hardening process is mostly complete. The egg has become a life container of amazing strength and resiliency with a nascent embryo inside.

For a time following fertilization—up to a month, depending on water temperature—the egg is vulnerable to shock by movement. During this period of sensitivity, eggs are never handled in hatcheries. Experience has shown that even with gentle handling, some eggs are certain to die.

The process of growth by cell division has long been known in general terms, yet its details remain both marvel and mystery. Thus it is with the velvety orange salmon egg secreted in the gravel. Having been fertilized, the egg cell begins the normal process of dividing and redividing until, on top of the internal egg mass, a tiny creature begins to form. This genesis always takes place on top of the sphere. If the egg happens to be turned, the yolk mass will rotate within the shell to bring the developing larva to the top again. In about a week, an embryo can be seen.

The first thing that can easily be recognized is the rudimentary eye; it looks like a black dot. Fish biologists refer to this phase of development as the "eyed stage." It is a preferred time for moving salmon eggs, such as in transplant experiments, to other watersheds. When the black pigment has appeared in the evolving eye, fish culturists know they can now handle the eggs with minimum risk.

Now the embryo can be seen moving inside its transparent capsule, dragging the yolk sac with it. Blood vessels form a lacy network over the surface of the yolk. This skein of blood vessels absorbs foodstuffs from the yolk into the bloodstream of the tiny being, discharging it into the main posterior vein below the now pumping heart.

The yolk—a rich mixture of water, fats, proteins, and salts—is the embryo's only food supply. From the yolk it receives both its energy and the materials needed for its growth. This food must last until the young fish leaves its gravel home and begins feeding on insect larvae and plankton. The period of internal feeding varies slightly between stocks and species of salmon, but by and large it extends from October to May.

INCUBATION

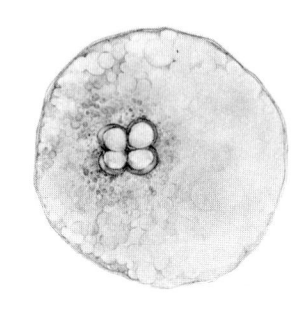

Embryo: four-cell stage

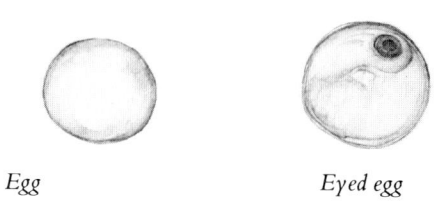

Egg

Eyed egg

Eyed stage

HATCHING

As the curled embryo grows and develops, its need for oxygen increases. For months it has nestled at the bottom of the streambed, hidden from view by the covering gravel, its life processes supported by the exchange of oxygen (in) and carbon dioxide (out) through the membrane of the capsule. But the time comes when the oxygen needs of the larva exceed the abilities of the membrane. About to suffocate, the larva begins to struggle. From a "hatching" gland on its head comes an enzyme which can dissolve the encapsulating wall. Agitated swimming movements distribute the enzyme. The rapidly thinning wall ruptures, releasing the larva—now an alevin—from its circular prison.

(If the larva is weak and fails to struggle during the hatching process, the enzyme will not be distributed within the capsule. In this case, the enzyme dissolves only a small portion of the egg wall. Through this hole protrudes the larva's head, its body trapped in the casing. Unable to escape, the infant fish dies in this position.)

Once hatched, the alevin discards the limiting membrane of the egg. Now it can absorb oxygen from the flowing water directly through its gills. The alevin can exist on half as much oxygen in its water supply as it required in the enclosed egg capsule. For this reason, if oxygen to the egg is reduced by low waterflow caused by silting of the streambed, early hatching can be a tactic of survival. Scientists have caused premature hatching of salmon eggs simply by lowering the oxygen content of the water.

ALEVIN

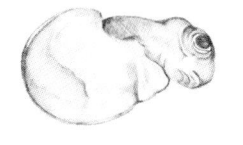

A hatchery is the only place where you can witness the birth of an alevin. Away from the hatchery, with its tanks and pumps and plumbing, alevins remain hidden in the gravel beds of spawning streams. They cannot be seen, but they are there. And they are dependent on the environment provided by that particular stream.

In nature, oxygen from the air dissolves into a water body at the surface. Water tumbling downhill through riffles and stones is thoroughly mixed and contains about as much oxygen as it can carry. In a river or a stream the lowest layer of water seeps through the gravel of the bed; as it does so, it loses its oxygen to bottom-dwelling organisms. To ensure the survival of salmon eggs and alevins, this inter-gravel water must be constantly replaced by oxygenated water from the surface, and it must be able to percolate through the gravel.

The rate of flow through the gravel depends on two things: the slope of the streambed and the porosity of the gravel. The steeper the slope, and the more open the gravel, the greater will be the flow. Usually, however, pebbles plug the spaces between the rocks and stones of the streambed. If sand and silt pack in around the pebbles, an aggregate can form which is so

Emerging alevin

Hatching

packed that waterflow almost ceases. Consequently, a direct relationship exists between the porosity of the gravel and egg/alevin survival.

When it hatches, the alevin reflects the shape of the egg: the yolk sac is still round and the alevin's body is bent. It resembles in contour the fetus of a human baby: all belly and head. Remarkably, the alevins hatch with the balance organ of the inner ear fully formed. They respond to gravity and at once assume an upright, horizontal position in the water. In a day or so the body stretches; the yolk sac becomes elongated. The alevins look more like the fish they are to become except, of course, you cannot see them. They live in watery blackness, hidden in the chinks and crevices of the streambed gravel.

Darkness is their refuge, and throughout this stage, young salmon avoid light. If suddenly exposed, they instantly dive for the shielding gravel. This is a survival reflex, for it is not until the end of the larval period, when the yolk sac has been consumed, that protective coloration appears. The young alevins are without countershading (dark back shading into a lighter belly) which would camouflage them; they have no barred sides and no reflecting pigment in their skin. Their bright orange yolk sacs make them easy prey for hungry older fish.

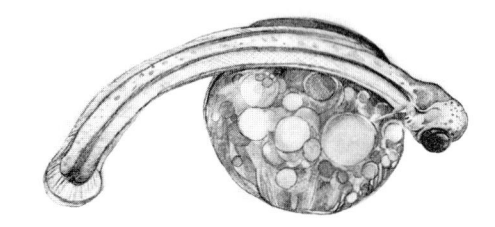

The physiological processes of baby salmon—heartbeat, breathing, digestion, and growth—all increase in warmer waters. Also, the young salmon need more oxygen. Assuming adequate oxygen, their growth rate will be determined by temperature and food. There are differences between races and species, but, in general, the rate of growth doubles between 0° and 5°C; and it doubles again at 10°C.

Emergent alevin

Since salmon eggs develop faster at higher temperatures, one might suppose that less yolk would be consumed by the alevin in warmer water, but exactly the opposite happens. At temperatures above 5°C, faster development is coupled with a reduced total growth. The result is that the fish (now an alevin) will evolve into a smaller than normal fry. This happens because in warmer waters metabolic processes are much less efficient.

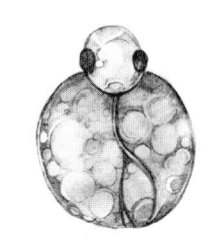

The alevin do not feed externally until they leave the gravel. Until that time they are wholly dependent on the nourishment in the yolk sac.

Head-on view

This fixed food supply must serve two quite different and conflicting demands: that for growth and that for metabolism. Whatever yolk must be used for energy needs cannot be used for growth; and if energy demands are intensified through increased activity, then growth *must* be reduced.

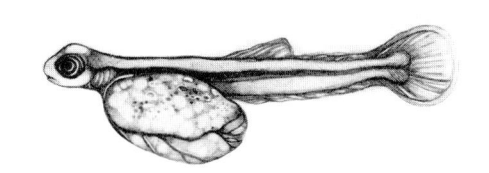

Alevin

FRY   When the yolk sac has been absorbed, the young salmon emerge from the gravel as fry. Two distinct guiding systems help them from the depths of the gravel into the open stream above: gravity and streamflow.

Under normal conditions, all that is required for the fry to leave the gravel bed is to move straight up, against gravity: its shortest way into the stream. But if the larval home is covered with silt or some impenetrable debris, the upstream response provides an alternative. Since the fish is already in a freshwater flow, passageways must exist in the direction from which the water is coming. This may not be so in a downstream direction.

There are many instances of salmonid fry appearing in ditches or wells hundreds of feet away from the spawning area. Such occurrences are explainable if we assume that these fry found their way to the surface blocked, switched to their alternative mechanism, and emerged at the origin of their water supply. On this basis one might predict that an underground waterflow will be found connecting the ditch or well and the natal stream.

Although it is hard to visualize young salmon or trout digging through the ground for considerable distances, the fish are well equipped for this chore. They have as yet no scales. Instead, their smooth tough skin is covered with mucus. Although they are capable of normal swimming, they have other methods of propulsion more suited to their gravelly surroundings. Swimming up a narrow crevice, for example, is done with a "trembling" of the tail fin, without undulation of the body as in normal swimming.

The fry can also move like a snake. By curving its body and pressing against the crevice walls on both sides, it slides itself ahead. To back out of an unsuitable passage the tiny fish can turn around in a surprisingly tight space. It can drop down passively or actively slither backwards. It can retract its head from too constricted an opening with a slow powerful tail beat. A sand barrier on top of the gravel is overcome by butting into it vertically; the sand grains drop past the thrusting fish, thus gradually opening up a passageway.

To this time the alevin has subsisted on its internal food supply. Once the yolk sac is absorbed, the alevin, now a fry, must reach the food sources in the flowing stream above. Now it leaves the sheltering gravel for a more vulnerable, free-swimming existence in the stream. Emergence of fry is also encouraged by warmer temperatures. For this reason large numbers of upward migrating fry approach the topmost layers of gravel late in the day. Here they will pause as long as daylight remains. With darkness, the inhibitory effect of light is gone and the miniature fish begin their travels.

When the young fry first leaves the gravel, it is still heavier than water.

Accordingly, its immediate goal is to reach the surface to inflate its swim bladder. The tiny fish swims upward, maintaining a nearly vertical position in the water. With a tail motion resembling vibration, it gains altitude in short stages. On breaking the surface, it snatches air with a sideways snapping motion of the head. The fry then drops back, keeping mouth and gill covers tightly closed. Some of the air is transferred through a duct in the upper gut into the air bladder. Several gulps of air may be necessary before the fish attains neutral buoyancy.

Each fish has to learn by trial and error the correct level of bladder inflation. Fry that have taken in too much air can be seen swimming head down to avoid floating to the surface. Tiny bubbles trail from their mouths as they expel surplus air. A few hours after emergence, all the fry will have achieved balance; all will be swimming in a normal horizontal position.

The typical daily "fry run" starts shortly after dusk, peaking about midnight. For the next few hours more surfacing fry appear, but as night progresses the activity slows. Sometimes there is a second peak of emergence shortly before dawn. Emergence is followed by immediate migration: stream-dwelling chinook and coho fry move to the shelter of the nearshores; sockeye start for their nearby nursery lake; chum and pink begin their downstream migration toward the sea. Normally, the fry run stops before dawn and does not begin again until the following evening.

FRY MIGRANTS

That salmon fry migrate has long been known. Over the years, biologists have studied the behavior patterns of the various species both in the lab and by field observation. A great deal has been learned, but there are endless exceptions and variations, and questions still remain.

Some species, or stocks within species, migrate *downstream* to reach their nursery areas. Some, such as sockeye or trout that spawn in lake outlets, must go *upstream*. Others, such as coho and chinook, may not have to migrate at all; the gravel from which they emerge is under their nursery stream. Some races of sockeye do not go to lakes; they spend their first year, or part-year, in river sloughs and estuarine backwaters. And although most Canadian chum go directly to sea, some spend days to weeks in their freshwater streams before leaving for salt water. (Some races of chum spawn on tidal flats where groundwater wells up. Their offspring emerge directly into salt water.)

Some chinook stay in fresh water for less than three months; others may stay a year. Some coho migrate upstream from their place of emergence; others migrate down. Many coho move out of the spawning stream into the river estuary, where they perish prematurely in salt water; others swim against the current, ascending to colonize nearby streams.

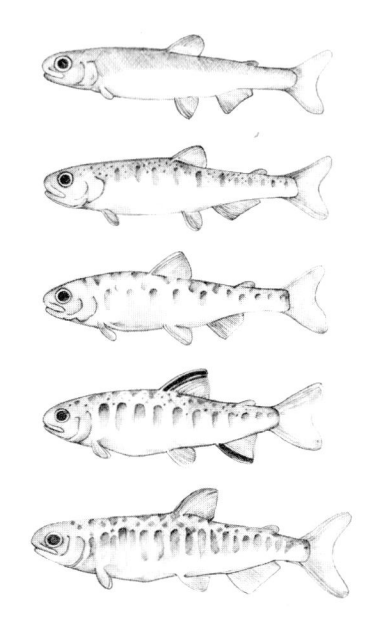

Salmon fry: pink; chum; sockeye; coho; chinook

Chinook, coho, and trout fry are stream dwellers. Their food consists of almost any organic matter, alive or dead, that drifts or swims past them. If there is insufficient food for the number of fish already there, newly emergent fry will be forced out by aggressive older and larger fry protecting their own territories. In many streams there is an annual emigration of hundreds of thousands of coho fry. For some, the exodus is downstream to the ocean. If the estuarine waters contain more than 20 parts per million of salt, all such coho fry are likely to die.

Since species, and even stocks within species, adapt to different freshwater habitats, simple explanations of fry behavior are not possible. Nevertheless, the lives of the five species of salmon fry can be described in general terms.

Sockeye remain in fresh water the longest. From their gravel incubation beds in the stream, the fry migrate to a nearby lake where they feed on plankton for one, two, or three years. In the late winter they begin a complex physical change, known as the smolting process, which adapts them to saltwater life. Outwardly, the skin pigmentation goes from a dull and barred appearance to a shiny silvery one. Inwardly, transformations occur that permit them to live in salt water.

Each spring when the ice has melted in the nursery lakes, the sparkling hordes of sockeye smolts move out, following the outflow rivers to the sea. When the surface waters of the lakes turn warm, the migration ceases. Any smolts that have not departed exchange their silver hue for the dull, barred exterior of a lake dweller. They will remain in fresh water another year. In the following spring, as water temperatures rise and the day length increases, these fish will again convert to smolts.

Migrating sockeye smolts

Coho stay in fresh water for one, two, or even three years. For the most part they are found in streams and the inshore areas of lakes. When small, their diet includes bottom-dwelling organisms and free-floating plankton. Later they feed on smaller fish and mature insects, primarily those falling from the air, trees, or banks onto the water surface. Coho also smolt and migrate in the spring, most of them entering the salt water in May and June.

Chinook live mostly in the larger river systems. A residence time of less than three months is common, but some stay longer, up to a full year. Like coho, as they leave the gravel of their birth, chinook fry spread out in both upstream and downstream directions, eventually to inhabit the accessible watershed. Both species feed by holding their position in a current and snapping up anything edible that drifts toward or past them. The chinook grow faster than coho, are therefore stronger swimmers, and tend to claim feeding territories in faster-flowing water. Chinook smolts arrive in the estuary over a more protracted period of time than other salmonids. They are, like coho, fish eaters, but they also eat plankton.

The chum migrate in spring directly into salt water where they remain in inshore waters for about three months. Some chum stay in the river for several weeks, but their number is negligible compared with those that move out within days. Chum are basically plankton feeders.

The species most divergent from the pattern is the pink. They depend the least on fresh water, move directly to the sea on emergence, and grow and mature the fastest. Along with chum fry, the pink immediately set out for the sea, alternately swimming and drifting with the current. If they cannot reach the estuary the first night, they dive for the river gravel, where they remain hidden during daylight. In large rivers, such as the Fraser, the migrating fry tend to school and travel downstream by night and day, depending on numbers (and muddy waters) to get most of them past the gauntlet of devouring predators. They continue this schooling tactic in the rich feeding grounds of the estuary.

Few salmon fry will survive to reach the open sea. Their populations are held in check by nature's controls: starvation, predation, and disease. In the lakes, streams, and inshore waters, predators contribute to the 80 to 90 percent loss of salmon during the fry and fingerling stages of their growth. The young salmon have two defense mechanisms: the panic response, in which they dart off in an unexpected direction, foiling the predator by the speed and suddenness of the move; and hiding. The hiding response extends over the salmon's freshwater lifetime when they seek the gravel crevices as newly emergent fry, hide in overhanging brush in pools and cutbanks as fingerlings, or, as adults, swim in the deepest parts of the river to avoid predators.

Food supply is a directly limiting factor, as can be seen by the territorial behavior of coho in streams. The larger coho fry aggressively defends as large a feeding area as is required to feed one coho; the size of the territory is determined by the food available. When supplementary food is added to the stream, the aggressive behavior declines, territories shrink, and the coho numbers increase dramatically. However, such increases brought on by man's interventions are likely to be temporary.

There is a second constraint on the numbers of fry a stream can support. Late in the fall the fish leave their territories to congregate in sheltered places for the winter. These are locations of quiet water under log jams, cutbanks, or up small protected tributaries. These sheltering areas are limited in number. Artificially increased summer populations of fry often do not survive until spring because there are insufficient wintering locales. This forces excess fish into premature migration to the sea where, it is believed, most soon die.

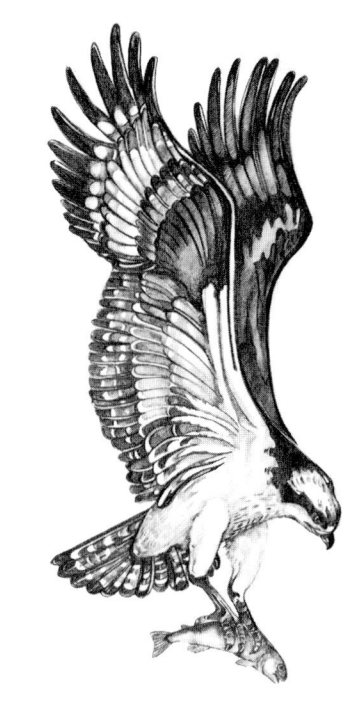

Osprey

ESTUARINE LIFE

Estuaries are the zones where the land and ocean processes interact. Material is carried from the land by the rivers and deposited in the estuary. The action of fresh water flowing into the seawater generates a circulation which brings minerals from the deeper water layers into contact with the surface layers, enhancing local primary production. Thus, freshwater outflow brings nutrients gathered from the deep ocean to the river mouth area. The result is a consolidation of nutrients from regions hundreds of square kilometers in extent into the sump-like estuary.

This wealth of nutrients supports teeming populations of microscopic organisms—phytoplankton and zooplankton—in both the upper layer of the estuary and on the surface of the intertidal flats and marshes. During the late spring, when the salmon juveniles begin their lives in the ocean, the production of plankton peaks.

Whereas the young of any of the five species of salmon or steelhead trout may be found in estuaries for roughly half the year, the time actually spent there by an individual fish is not known with any precision and is thought to be quite variable. Results of tests using marking and recapture techniques suggest that two weeks is a reasonable estimate. In these two weeks the fish adapt to seawater and grow at an astonishing rate; they may double or even triple in size.

This spurt of growth occurs at a critical phase in the life of the fish. Being eaten is probably the greatest single cause of mortality of juvenile fish of all species. One of the best protections against predation is rapid growth, and for the first two weeks of their marine life the young salmon appear to be doing just that: escaping into larger size.

OCEAN DISTRIBUTION

From tagging studies it is known that some juvenile salmon remain for a brief period in the estuary of their home stream and then commence their ocean migration. The extent and direction of their travels vary, depending on time and size when they enter the ocean, marine conditions, and other factors.

Stocks of Pacific salmon and steelhead trout leave the coasts of British Columbia and southeastern Alaska, then turn north to travel along the coast in a narrow band extending about 40 km (25 mi) offshore. This movement continues around the periphery of the Gulf of Alaska northward and westward past Kodiak Island. The band widens in the northern part of the Gulf, presumably because of the widening of the continental shelf. Stocks of young salmon from Prince William Sound, Cook Inlet, and Kodiak Island join the procession from southeastern Alaska and British Columbia.

Between fall and midwinter, Pacific salmon juveniles leave the coastal belt to make a continuous wide sweep around the rim of the Gulf of

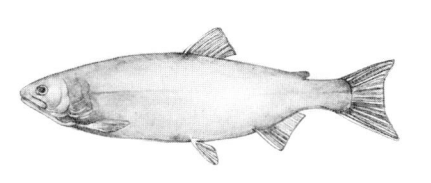

Sub-adult

Alaska. The migration takes them hundreds of miles to the south, where they scatter widely over their winter feeding grounds.

There are exceptions. Some Columbia River stocks have been found to turn south as they leave the Columbia and are caught in Oregon and California waters. Marked fish released from the Capilano and Big Qualicum hatcheries in British Columbia have been recovered as far south as Coos Bay, Oregon.

The distance a salmon will travel varies as much as the direction. Some hatchery-bred chinook stocks migrate over long distances, spending much of their adolescent life in the Gulf of Alaska before returning to begin a long freshwater migration into the upper Columbia or upper Fraser river systems. In general, coho tend to migrate shorter distances than chinook, if only because they are at sea for a shorter time.

There are stocks of both chinook and coho that are known to stay close to home. For instance, some stocks spend their entire marine lives in the protected waters of Georgia Strait and Puget Sound and do not migrate to the open ocean. These fish tend to be smaller at maturity than fish that have gone "outside." They are important, however, for they provide the basis of a year-round sport fishery.

At present little is known about the ocean migrations of steelhead trout. There is no ocean sport fishery for this species and those caught commercially are taken incidentally in the salmon fishery by gill nets or seines. Future tagging programs may yield new facts on this important sports fish.

HOMING

The stocks of salmon from many rivers intermingle in the oceanic pastures during much of their marine lives. As they approach maturity they begin a movement to coastal waters. Close to land they disperse, each salmon seeking its parent stream.

Proof that salmon return to the freshwater breeding grounds of their origin has taken years to accumulate. Early evidence came from marking seaward migrating juveniles and noting the recovery in the same streams of adult fish bearing the same mark. More recently, scientists learned that biological characteristics such as scale markings or specific parasites can distinguish fish associated with certain watersheds from others. Salmon tagging programs have been conducted from research vessels far out in the Pacific Ocean. These tagged adult salmon when caught in the inshore fishery enable researchers to estimate the time and the distance they have traveled.

We now know that the shoreward migration of maturing fish from the high seas is not a random occurrence; it is well timed and well directed. In predictable order the different stocks and species of Pacific salmon aban-

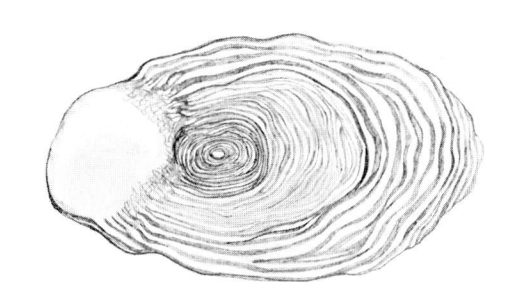

Salmon scale with growth rings

don their ocean feeding grounds to travel to their different spawning locations.

During their saltwater trek, juvenile salmon from British Columbia travel at about 20 km (12 mi) per day to reach their wintering grounds. This is slow compared to the homeward speeds of adult salmon. Karaginski pink homing from the central Aleutians average 50 km (30 mi) per day during their final days at sea. Bristol Bay sockeye do the same. Skeena River sockeye average 36 km (22 mi) per day during their last month of ocean life. These speeds are calculated on the shortest distance between tagging and recovery.

More precise methods have been used. Sockeye tagged with ultrasonic "beepers" off the northern British Columbia coast averaged 2 km per hour. In laboratory experiments, adult sockeye have been found to swim without rest for at least 100 hours at 2.2 km per hour. To cover distances of 30 to 50 km (18 to 30 mi) per day—while traveling hundreds of kilometers of Pacific Ocean—salmon must swim almost 24 hours a day.

Mature sockeye returning from the sea

Such a physical feat is astonishing, yet it is but a prelude to a more arduous journey: the ascent through the freshwater streams to the ancestral spawning grounds. Some stocks of salmon travel thousands of kilometers through rivers, lakes, and tributary streams before concluding their reproductive odyssey. An example is the Yukon River chinook run. After entering the river from the Bering Sea the fish travel some 3300 km (2,000 mi) to reach their spawning grounds on Michie Creek in the southern Yukon Territory.

FISHERY

Migration slows and often stops when the salmon reach estuaries or coastal waters. Sockeye tagged along the west coast of southeastern Alaska traveled the greater part of the journey to the Skeena River at an average rate of 37 km (23 mi) per day. Close to the river's mouth, the run slowed to a rate of 6 km per day. The same behavior applies to pink salmon on their way to the Fraser River. This migratory behavior may be a general characteristic of all maturing salmonids approaching their home estuary.

In coastal waters waits man. Commercial fishermen use gill nets which snare their catch, seine nets which encircle it, and troll gear which lures and hooks the fish. Indian food fishermen use gaffs, spears, dip nets, and traps; sports fishermen use hook and line. Of these methods gill nets are the most widely used.

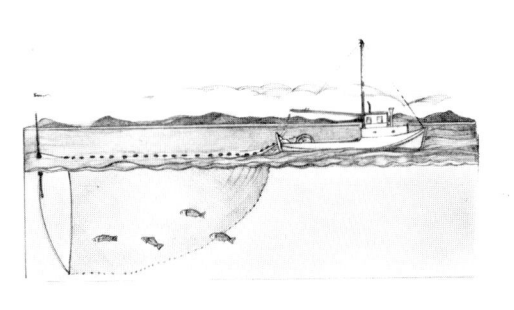

Gill-netters are usually one-man boats equipped with a power drum mounted near the stern. The fishermen chooses a location through which the salmon are migrating and lowers his fence-like net across their path. Salmon become snared in the mesh barrier when they try to swim through

Gill-netter

45

it. Nets are dyed to reduce their visibility in the water. Also, gill-netters take advantage of darkness to make their nets less visible to the fish.

One end of the net is attached to the boat, the other to a buoy. Both net and boat drift with the tide or current from one to four hours, depending on location, tides, and catch. The fisherman recovers his net by rolling it onto the power drum; as it is pulled in over the stern, he extricates the salmon from the net.

Since a gill net acts as a snare, the size of the mesh determines the size of the fish caught (and thereby species and sex). The fisherman generally has a number of gill nets of different sizes and colors so that he can fish for various species and sizes, and in locations of different water color. All salmonid species are caught by gill nets, but major landings are of sockeye, pink, and chum salmon.

Setting of gill net

Farther offshore the commercial troller, poles extended from either side, trails fishing lines festooned with lures and hooks. The lines are fed out through pulleys on the ends of poles and down into the water. A "cannonball" weight holds the line to a selected depth. The jerking of a fish caught on any lure rings a bell on the end of the pole, indicating to the fisherman which line to pull in.

Until recent years chinook and coho were the main species caught by trollers. With new types of lures, sockeye and pink are being taken. Another development is the combination troll and gill net boat.

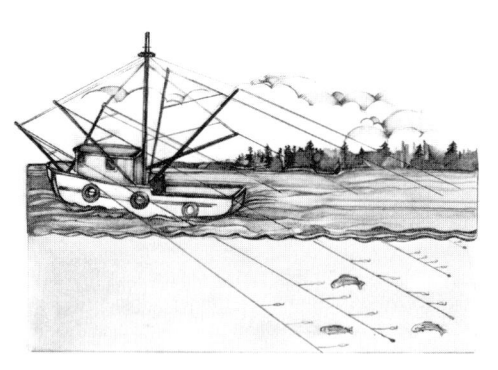

Salmon purse seiners are four to eight-man vessels, equipped with either a boom-mounted power block for net hauling or, more commonly, a power drum near the stern. The net is set over the stern with the free end attached to a skiff, drogue, or buoy, or to a fixed object on the beach. The net is supported in the water by floats along the top and is kept vertical by lead weights along the bottom edge. A purse line passes through rings attached along the lead line.

Troller

When the seine is set in a large circle around a school of fish, the two ends of the seine are brought together at the boat and the purse line is tightened to draw the bottom of the net together, forming a "purse" in which salmon are trapped.

Either a power block or a stern-mounted drum is used to haul in the net until fish are massed in a small section of the net alongside the vessel. The fish are scooped aboard with a dip net, or "brailer." When there are few fish, the purse and its thrashing silvery contents are sometimes lifted directly on board.

All species of salmon are caught by seiners, but the catch is primarily pink, sockeye, and chum.

The recreational salmon fishery takes mainly chinook and coho. These species are caught with natural food baits, mainly herring, or with artifi-

Purse seiner

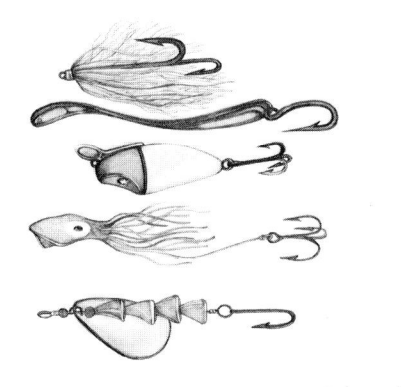

cial lures which imitate natural prey. Salmon are fished for sport by trolling from a moving boat, casting and retrieving, or by still-fishing or "mooching." Sportsmen may take chinook, coho, and steelhead in the sea and in freshwater streams, but sockeye, pink and chum must be released in non-tidal waters.

Indians of the Pacific Northwest capture salmon for their own consumption by many methods, including gill nets, set nets, gaffs, and traps. A set net is simply a gill net attached to a fixed object so that it does not drift with the current or the tide.

A gaff is a steel hook on the end of a pole. The gaff is moved through the water, usually where fish gather below a river obstruction. When the fisherman feels something bump the pole, he jerks it up, impaling the fish on the gaff hook.

Dip nets are also used in places of river constriction. In this method the fish are trapped in a small net bag instead of being impaled. In other areas, Indians still set traditional weir traps or fish with spears.

By far the greatest number of salmon are caught by commercial fishermen. Forgotten these many years are the "rowboaters," or hand trollers, who plied the inshore waters of the British Columbia coast by the hundreds. They worked alone in their double-enders, at the oars off and on from first light to late dusk. In a good season—five months, usually—a man or woman might make $300.

The single-line hand troll fishery reached its climax and ended in the 1930s. With the coming of World War II there appeared easier ways of making a living than selling a day's catch of hand-caught salmon for two or three cents a pound.

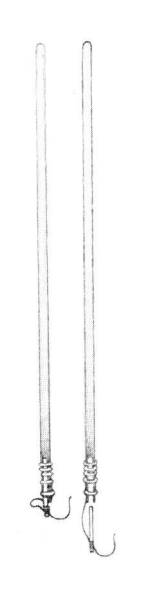

Salmon lures

Gaffing

Gaff

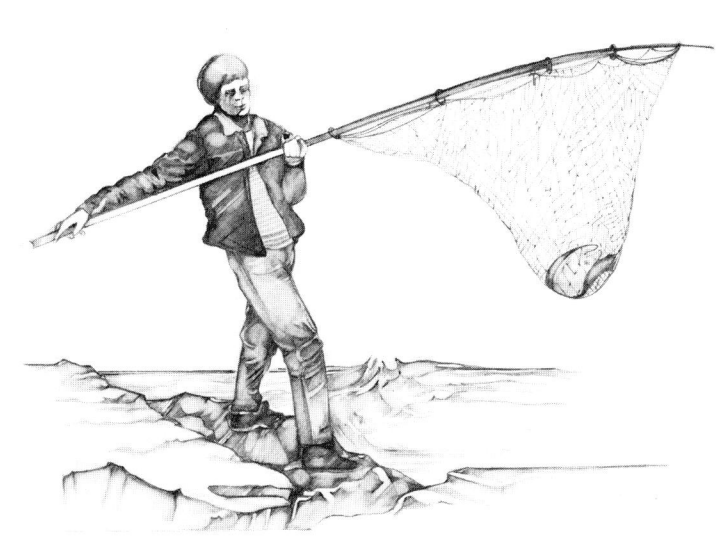

Dip netting

Environment

The outset of the 1970s saw the beginning of what became known as the Environmental Crisis, when people everywhere were suddenly alarmed at the general deterioration of the world around them, and pollution became a major issue. Conservationists found militant allies in the environmentalists. The movement developed its own lexicon: population explosion, recycling, endangered species. It was a time of new concepts: of Spaceship Earth, the idea that we are all passengers aboard a ship whose life support sytems are being sabotaged by the crew; that man's impact on other life forms resembles that of an epidemic disease; that overkill methods of hunting and fishing have destroyed entire species; that industrial chemical processes, monoculture farming, deforestation, radiation, and acid fallout have taken a murderous toll of the world's creatures.

The Pacific coast of North America has all these problems. Some are not new. By the middle of the twentieth century much of the damage to salmon streams had already been done by urban encroachment, road and rail construction, logging, and other industrial activities in the salmon watersheds. Entire fish stocks had been destroyed through the combination of decreasing stream production and increased fishing pressure. By the 1950s many west coast streams had only vestiges of their salmon runs.

In British Columbia, industrialization began at tidewater. Pulp and paper mills are an example. For decades the tidewater mills pumped their dissolved chemical wastes directly into the sea. These chemicals were highly toxic to fish, but before environmental legislation to control these discharges was enacted, the pulpwood supplies near the coast had declined. A second generation of mills were built in the hinterlands, but even though these modern mills provide retention ponds to oxidize chemicals in the waste water before it is returned to the nearby stream, the mills continue as a threat to salmon.

The wood wastes poured into the smaller inland rivers have an enriching effect. With enrichment comes a process of eutrophication. In this process algae proliferate, using nutrients provided by the organic materials in the water. Mats of algae form on the bottom of the streams, interfering with the flow of water through the gravel.

Wood fiber lost in the pulping process is also deposited in the streambed. As it decays, it consumes the oxygen needed by organisms living in the stream bottom. A mill producing 300 tonnes (350 tons) of bleached kraft pulp per day can lose one tonne of fiber daily in its waste water. Streams so abused are useless to salmon for spawning or for egg incubation.

Other manufacturing and agricultural operations discharge nutrient wastes into the water. These sources include drainage from feedlots and pastures, process wastes from fruit and vegetable canning and starch manufacturing plants, packing house wastes, and fertilizer drainage from farms and orchards. The degree of enrichment depends on the amount of dilution afforded by the affected waters.

As eutrophication progresses, algae and rooted aquatic plants flourish. These plants die, ending their natural cycle of growth, and eventually blanket the stream bottom with decaying organic matter. Again, oxygen is robbed from the water, with insufficient amounts remaining for the development of salmon eggs.

Another threat to salmon is domestic sewage from urban centers. Initially, when communities do not have much sewage, they flush their wastes directly into the nearest watercourse. But with urban growth, the need for at least primary treatment—the simple removal of solids—is acknowledged by most people. Processing beyond this stage is more costly and, therefore, more controversial.

Secondary treatment reduces the oxygen consuming capacity of the waste water and converts ammonia and other toxic materials to less poisonous forms, such as nitrates. However, such treated effluents still contain large amounts of nitrates, phosphates, and detergents, all of which act as nutrients. Removal of these fertilizers requires tertiary treatment at an even greater expense to the community.

How much sewage treatment should be undertaken is usually dictated by its cost. Against this dollar cost must be reckoned the potential ruination of the freshwater body for other uses, especially the loss of spawning grounds and the commercial and recreational values of the self-perpetuating salmon resource.

Each year from early spring to late fall the rivers that drain the Pacific West Coast are filled with migrating salmon, the young migrating to the sea, the adults returning to spawn and die. The success of those journeys depends on the conditions of the freshwater streams.

Each new generation of salmon begins life in a gravel nest in the streambed where the spawning female deposited her eggs, then covered them with clean gravel. The stones protect the eggs from harmful light and from hungry gulls, ducks, and trout.

There are other dangers. In their earliest stages of development the eggs can be injured by disturbance. Other spawning salmon may dig up the eggs, some of which will be eaten by waiting predators. Some will drift back down into the gravel, but of these, many will have been damaged and die. Fungus growths develop on them and spread to healthy eggs nearby. The fungus eventually covers the eggs and, by interfering with water circulation, suffocates them.

Suffocation can also be caused by sand and silt sealing the gravel during autumn rains and high runoff. The cascading waters may even move the gravel downstream, damaging the eggs or washing them away. In winter, gravel beds may freeze, including the salmon eggs they contain; in the spring, after hatching, the alevins may be flushed out of the protective gravel before they can fend for themselves.

These threats exist under natural conditions; the activities of industrial man magnify the threats. In times past, road builders have used streams as quick sources of clean gravel. Its removal destroys the spawning ground, as well as the eggs or alevins it conceals. If the eggs and alevins are not crushed by heavy machinery using the streambed as a roadway, they die when the gravel is compacted, slowing the life-sustaining waterflow through it.

The aftermaths of logging, forest fires, and the clearing of land for agriculture are equally pernicious. Removal of the forest cover immediately accelerates the drainage of rainfall and snowmelt from the land. Watercourses become prone to flash floods that expose and carry the subsoil into the streambeds. The problem is aggravated by road, rail, pipe, and power line construction. The gravel bottom becomes filled with sand and silt, so that the stream is ruined as a spawning area.

In earlier years, river driving of logs was a standard practice, and the results are still with us. Logjams, and logging slash carelessly dumped, forced streams out of their normal channels. When this happened spawning beds were cut off and became dry. In hot summers, the streams became low or dried up completely. Where shade trees vanished because of farming, or forest fire, direct sunlight raised water temperatures too high for the salmon.

Another problem afflicts rivers once used for transporting logs downstream. Bark from log drives of former years accumulated in eddies and backwaters, where it sank. It continues to decay and, in decomposing, extracts oxygen from the water. Any streambed blanketed by chips and bark has been ruined as a suitable home for spawning salmon.

Fire aftermath

ENEMIES

After hatching, the alevins remain in the gravel, sustained by the food energy stored in their yolk sacs. In March, April or May, when most of the yolk is absorbed, the young fish work their way to the surface of the gravel to emerge as free-swimming fry. At once they start for the nearby feeding grounds. Their enemies are numerous and are everywhere. They include parasites, predators, and disease.

The parasites come with the plankton organisms they feed on. Many of these take up residence in the young salmon's gut, in the body cavity, or in the flesh. Others attach externally to the skin or the gills. Parasites slow the growth rate of the young fish which, being smaller, are vulnerable over a longer period of time to predation.

In the short journey from natal stream to their nursery lake, sockeye fry run a gauntlet of waiting sculpins, trout, char, whitefish, and squawfish. Those that reach the lake remain prey for these and other fishes, including freshwater ling and larger coho juveniles. Diving birds such as ducks, grebes, loons, mergansers, and kingfishers also take their share.

Predation is linked to food supply. Sockeye fry rearing in lakes, for instance, compete with others of their species, and with whitefish, sticklebacks and minnows, for the available food supply. When the plankton supply is limited, their growth is retarded. Smaller for a greater length of time, the fry remain susceptible to the predation they would outgrow in more favorable years.

The fry of pink and chum school and feed in the rich coastal waters. With food supply assured, they grow rapidly. They are exposed to predation from many types of marine fishes and diving birds, but for a shorter time.

FOOD

Natural calamity has never equalled man's interference with the freshwater habitat of the salmon. The removal of forest cover results in flash floods and freshets, whose torrents of muddy water scour the stream gravels, flushing bottom-dwelling organisms out of the nursery streams. In summer the streams may slow to trickling rivulets. Any wide fluctuations in flow will change the pools and riffles so important to the fry. The pools provide cover for the young fish, who feed on organisms drifting down from upstream riffles. When the gravel is alternately dry, then disturbed by freshets, bottom organisms lack the stable conditions they need to live and propagate. As a result there is simply less food available for the resident fry.

Salmon are cold-blooded animals, and the amount of food they require to sustain themselves increases with rising temperatures. Above 15.5°C the fry may have difficulty finding enough food. At the same time, their resistance to disease falls off. Around 24°C the young salmon will die from temperature alone. More frequently, however, streams dry up before they become too warm.

Other man-induced hazards exist. Nursery streams are sometimes hit with forest sprays used to control insects, or with defoliant sprays used to check unwanted shrubbery growing along stream banks. Such chemicals can kill young salmon, or they may be indirectly harmful by affecting food organisms living in the stream.

Nursery lakes suffer two major abuses from man: the dumping of wastes, and changes in water levels. The first of these can include domestic sewage, mine tailings and chemicals, industrial effluents, fertilizer and pesticide drainage from cultivated land. These substances may either be poisonous or they may enrich the lake waters, causing eutrophication.

Changes in water level, on the other hand, begin when lakes are impounded and the lake level fluctuates with dam storage needs. Severe damage occurs in the shoreline, the part of the lake that produces most of the food for the young salmon. When the shoreline is permanently flooded—or alternately flooded and left dry—much of the productivity of the shoreline shallows is lost.

DAMS Of the hazards that face seaward migrating smolts, dams are the most serious. The first obstacle is the slack waters of the impoundment. The smolts, confused by the lack of current, gather in the deeper waters. Some are eaten by resident fish predators; others abandon the migration and live out their lives in fresh water.

Most, however, are driven to finding the sea. Confronted with a hydro power dam, the smolt has a choice of two ways to continue: through the deeper turbine intakes, or over the spillways. The greater number try the turbines. The danger is death or injury from the spinning turbine blades, or from extreme changes in pressures as they plummet through the tubes to emerge below the dam. Smolt losses going through the turbines are between 10 and 40 percent.

The other route—over the spillway—is no less hazardous. Depending on the spillway design, smolt mortality ranges from 5 to 70 percent. The losses due to turbines and spillways are appalling; nevertheless, the ordeal for the survivors may not be ended.

As the water pours over the spillway it entrains air. If the height is great, this entrained air dissolves under the increased pressures of the depths below the dam. The water becomes supersaturated with nitrogen and oxygen. Under these conditions smolts may die from the "bends": gas bubbles forming in the blood and tissues.

On the West Coast, because of the mountains, high dams can be constructed for hydropower generation, and these are often built in series. Power dams and salmon cannot coexist on the same rivers. In the past, governments have refrained from building dams on major salmon rivers in British Columbia; instead, as electrical power needs grew, dams were installed on some of the non-salmon rivers and on salmon rivers of lesser importance. In years to come, however, the question of salmon versus dams may have to be reviewed. The decisions will not be easy ones. They involve chipping away at the edges of the main salmon populations through the damming of minor rivers; or damming large salmon rivers and decimating major salmon populations; or finding alternative sources of power.

Thus far, British Columbia has resisted the temptation to dam the Fraser River which traverses the province in a 1400-km (900 mi) long S. In addition to being the largest salmon producer on the West Coast, the Fraser could be one of the largest power producers. The series of mainstream dams that would be needed to realize the power potential of the Fraser would alter the watershed forever and wipe out every important salmon run. The argument between the developers and the conservationists has gone on for the past 30 years: power versus fish.

In 1955 B.C. Hydro, the provincial government-owned utility, proposed a 230-m (750-ft) high dam across the Moran Canyon above Lillooet. It was not until 1971 that the Moran Dam proposal was finally shelved as a result of an adverse report by the Canadian Fisheries Department. Less than three years later B.C. Hydro was proposing the McGregor Diversion Project as the first step in a plan that included five power and storage reservoirs and three run-of-the-river hydro plants in the Fraser system. Although public pressure succeeded in shelving the project in early 1978, the scheme serves to illustrate the problems.

The McGregor River drains a portion of the Rocky Mountains, joining the Fraser as a tributary some 80 km (50 mi) northeast of Prince George, British Columbia. By diverting the McGregor's flow from the Fraser to the northward-flowing Parsnip River, the power company planned to add to the head of water in Williston Lake, which is the reservoir for the W.A.C. Bennett Dam. It is a low-cost way to boost the dam's power output.

The diversion was expected to lower the flow of the Fraser by some 6,000 cubic feet per second or by about 20 percent of the total flow at low water times. A Canadian Fisheries Department study estimated that the effect of this single alteration would be to wipe out 1.7 million salmon a year. But it was only an estimate, based on the knowledge then available.

As the rate of flow decreases and the levels fall, the sun has a better chance to warm the remainder of the water. The effects of higher temperatures on chinook and coho are not well known, but the lower water levels would be damaging, for, though the offspring of sockeye rear in nursery lakes, coho and chinook fry grow to seagoing smolt size in calm areas of the river itself. Less water in the Fraser would mean less rearing space.

There is yet another hazard to salmon from the proposed McGregor River diversion, that of the introduction of northern pike to the Fraser watershed. At the moment, this fierce freshwater predator is confined to waters flowing eastward from the Rockies. With the McGregor dam in place, the pike might make its way from the Peace-Parsnip easterly drainage system into the Fraser (draining into the Pacific). In addition to preying on salmonid fry and competing for food, the pike is host to the parasite *Triaenophorus crassus*. This tapeworm could lower the commercial value of the salmon; at worst, it could be an ecological tragedy.

The 140-m (460-ft) high proposed McGregor River dam was originally presented to the public as a means of preventing flooding along the lower Fraser. As mentioned earlier, it was but the first of a complex of dams and reservoirs intended for the Fraser River. B.C. Hydro had, as of April 1976, 23 dams, 18 power plants, and 11 reservoirs in the active planning stages throughout British Columbia. To the northeast, the Peace River will have two more power facilities built between the W.A.C. Bennett Dam and the Alberta border (with calamitous effect on the downstream Peace-Athabasca Delta).

Salmon rivers in British Columbia slated for power generation development include the Homathko, Stikine, and Skeena. There are five projects planned for the Homathko, with the diversion of water from the Taseko and Chilko tributaries of the Fraser. At risk here would be the famous Chilko River sockeye run.

There are five potential power projects listed for the Stikine River and two on its tributary, the Iskut. B.C. Hydro's Cutoff Mountain project on the Skeena River would be 13 km (8 mi) upstream of its confluence with the Babine River. The 260-m (850-ft) high dam would threaten the valuable salmon runs of the Skeena.

To avoid this, British Columbia may turn to nuclear power plants, but when these are built on rivers and estuaries yet another hazard to migrating salmon will be introduced: that of thermal pollution. Nuclear plants use large amounts of water for cooling purposes. When the heated effluent is returned to the adjacent river, lake, or estuary, the effect on the ecosystem is almost always injurious. For salmon passing through these zones of heated water, the effects are certain to be adverse.

POLLUTION   High dams and nuclear power plants are easily seen as threats to salmon and trout. Less visible are the sublethal effects of toxic chemicals that find their way into the lower Fraser River. Compared to other heavily used waterways in the world, the water quality of the lower Fraser is good. Nonetheless, abnormal quantities of toxic substances, such as mercury,

The Catch and the Environment

are found to be accumulating in crabs and other bottom-dwelling organisms of the river. No one knows whether seaward migrating salmon smolts on encountering a toxic discharge are affected by the poison and made more vulnerable to predators waiting in the nearby estuary.

The Fraser estuary resembles that of others around which great cities have arisen. Such cities have tapped groundwaters for use but have returned the wastewaters to the sea instead of to the land. For this reason pollution problems are the most acute in the harbors, estuaries, and coastal waters of the world. Toxic heavy metals precipitate from industrial wastes and usually remain in the estuary (as do DDT and PCBs). Dredging spoils and landfill operations gradually eliminate the bottom life of the urban estuary. The productivity of the Fraser estuary has declined by an estimated 70 percent. In a few decades it may be destroyed if present pollution levels continue.

Industrial liquid wastes are the largest source of pollution in coastal and estuarine regions, followed by municipal sewage. The runoff from the agricultural land of the Fraser Valley includes manure, pesticides, and fertilizers, and these further pollute the river and estuary. Sewage and fertilizers can so reduce oxygen in the water that fish and other aquatic species cannot survive. Unfortunately, the red, black, and other colored "tides"—population explosions of micromarine organisms, or "dinoflagellates"—thrive on the sewage being dumped offshore by world cities.

These microorganisms have the ability to move like creatures but, like plants, utilize sunlight to make food. When triggered by favorable conditions they begin multiplying until there are millions per liter of seawater—in such numbers that they appear to stain the water. Some dinoflagellates are poisonous, killing off fish and other marine creatures. Others accumulate in shellfish without killing them, but are lethal to humans who eat the shellfish. When the blooms die off (as suddenly as they appeared), they create low-oxygen areas of water in which other forms of marine life are suffocated.

Pollutants enter animals, including man, through biological concentration. Billions of microscopic phytoplankton act as a great biological blotter, picking up nutrients, trace metals, PCBs, and other materials. Zooplankton feed on the phytoplankton and successively pass the pollutants on to higher organisms. As this process moves through the food chain, concentrations reach their highest in such predators as marine mammals, birds, and man.

The lethal and sublethal effects from toxic wastes are complex and only partially understood, but the evidence is mounting that these effects may be widespread and harmful to ocean life. Man is changing the chemical

Shuswap Lake

Alberni Inlet

Indians food-fishing for salmon in Stamp River

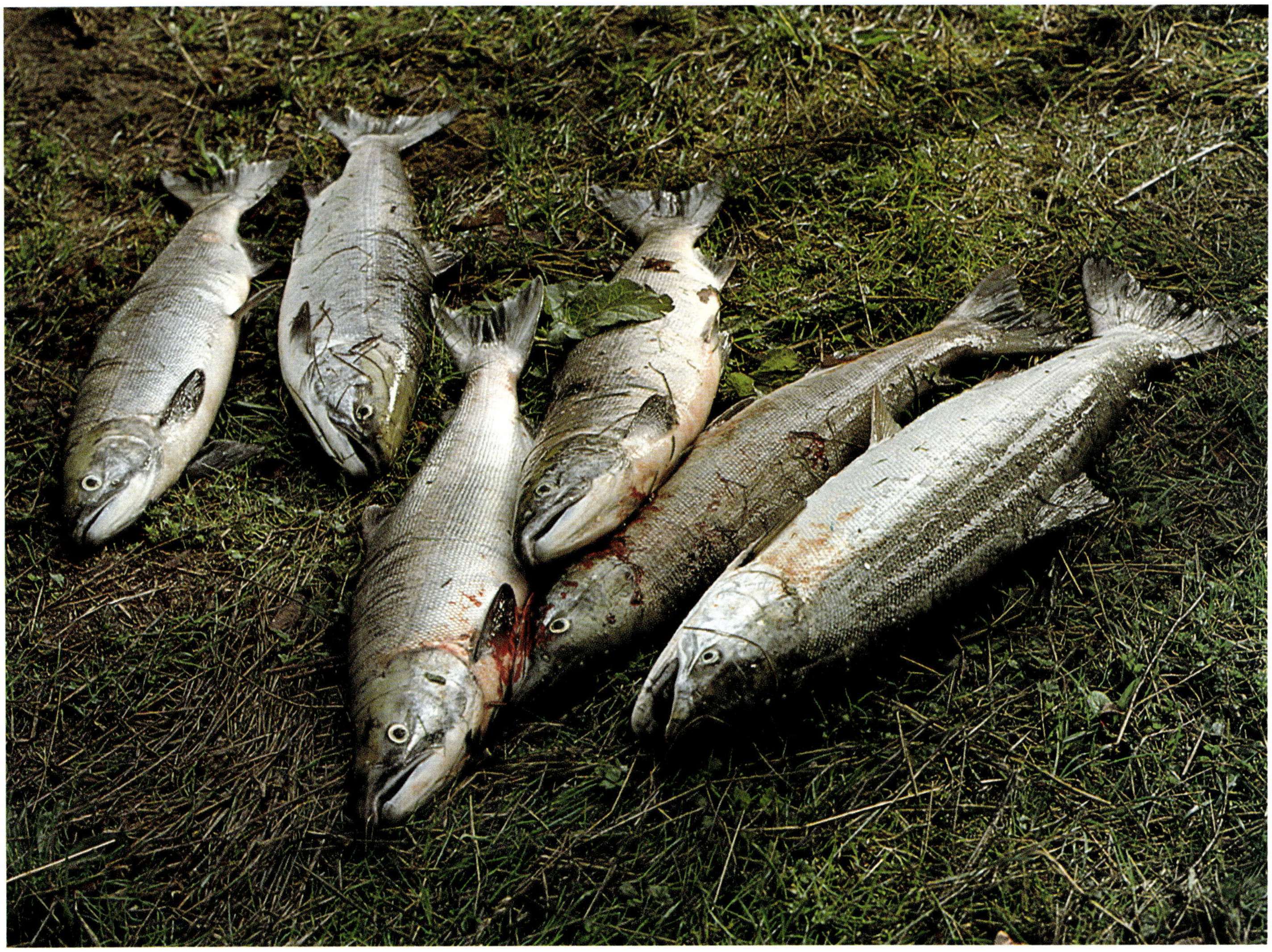

Sockeye wind-drying prior to smoking

Roasting salmon

Indian smoke house along Babine River.

Woman of Babine Lake Band filleting sockeye

Sportfishermen take 13 percent of annual salmon catch

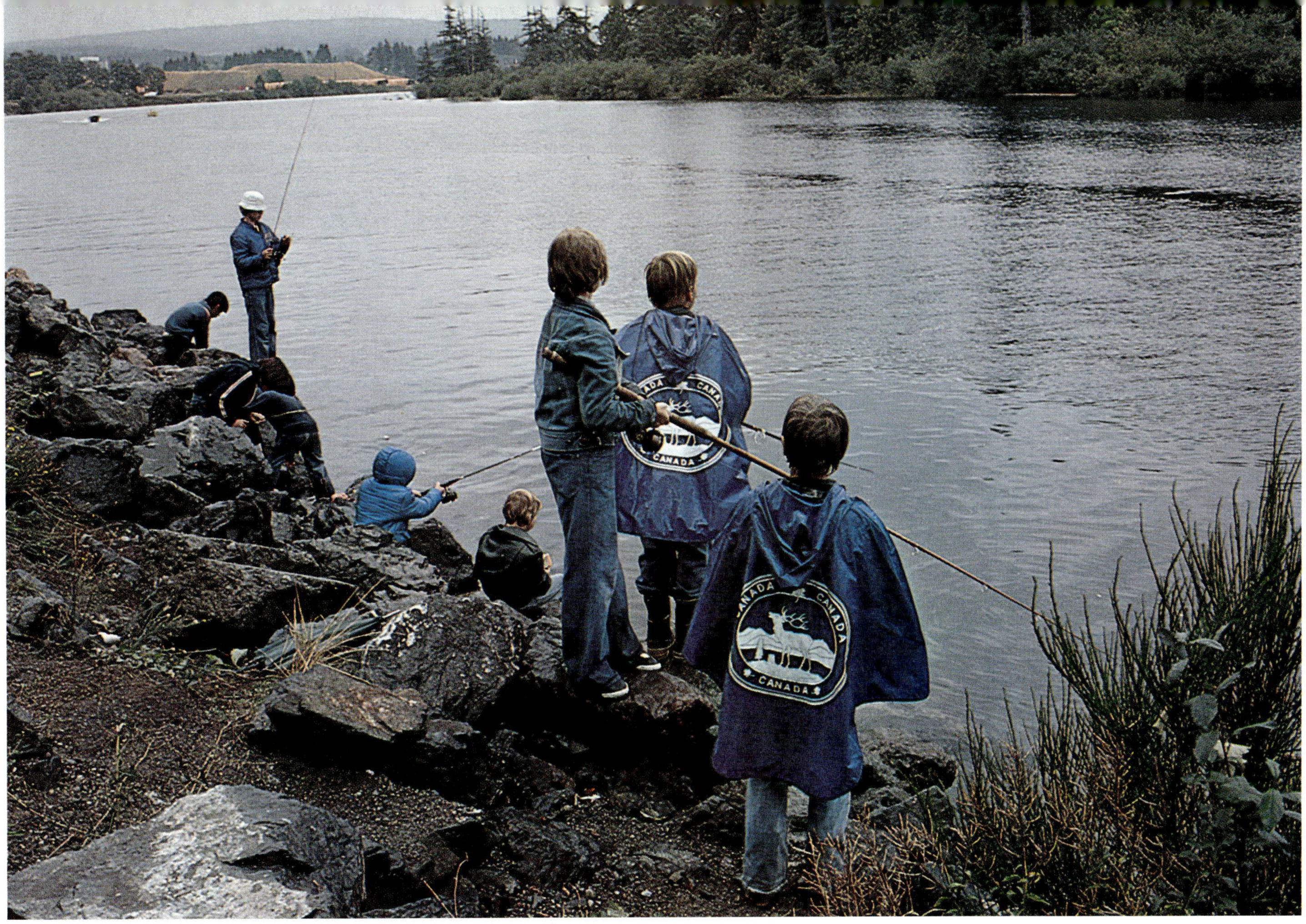
CANADA
CANADA
CANADA

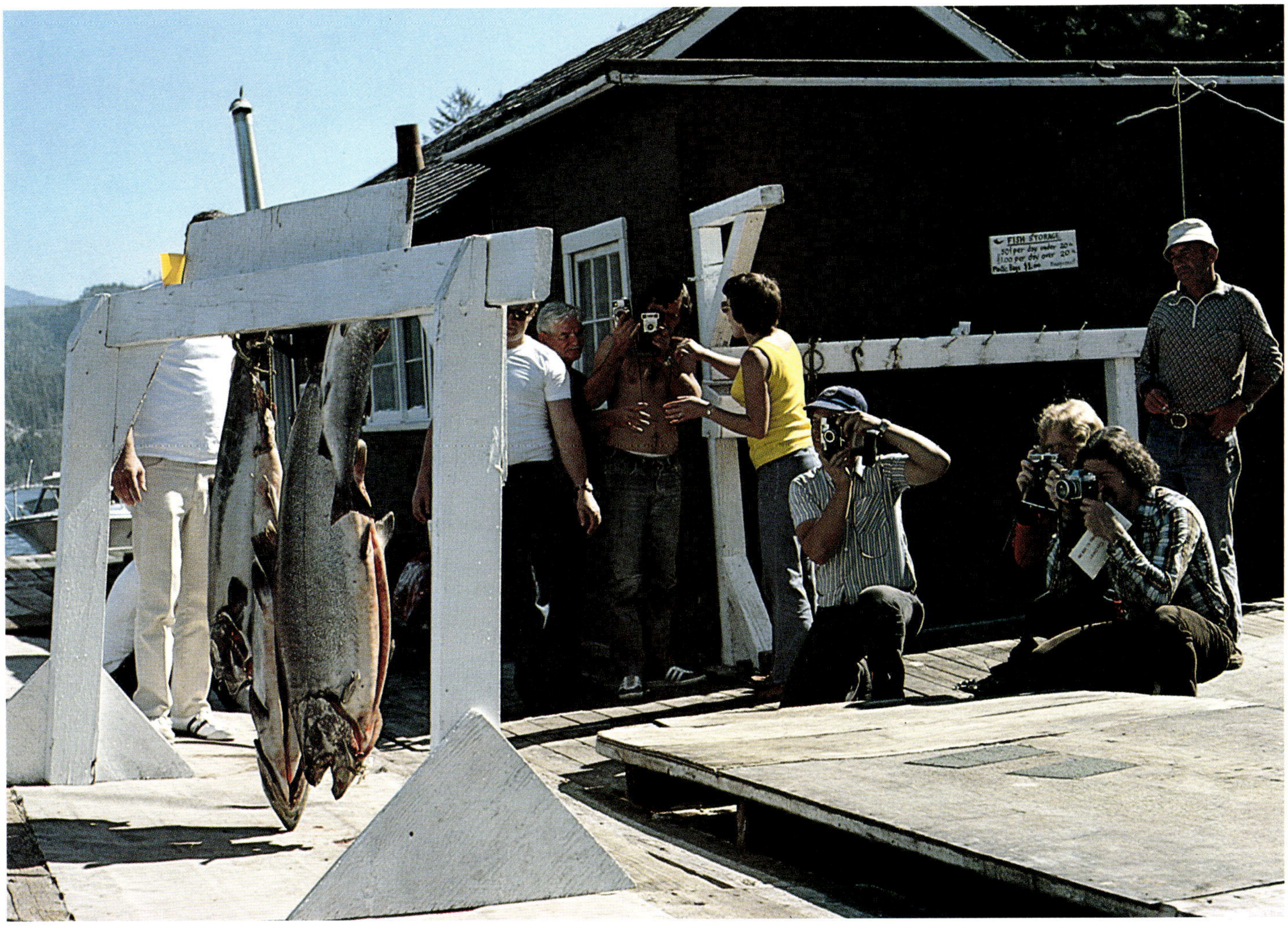
FISH STORAGE

Commercial fishing is also done from small boats

Salmon seiner in Barkley Sound

Seiner in Queen Charlotte Strait

Seiner set

Longliner at Bamfield

By autumn salmon eggs are incubating in streambed gravel

Flowing water brings oxygen to eggs

In spring juvenile salmon head for salt water

Aerial fertilization of Great Central Lake

Clear water from the Thompson River joins the muddy Fraser

Stewart Lake

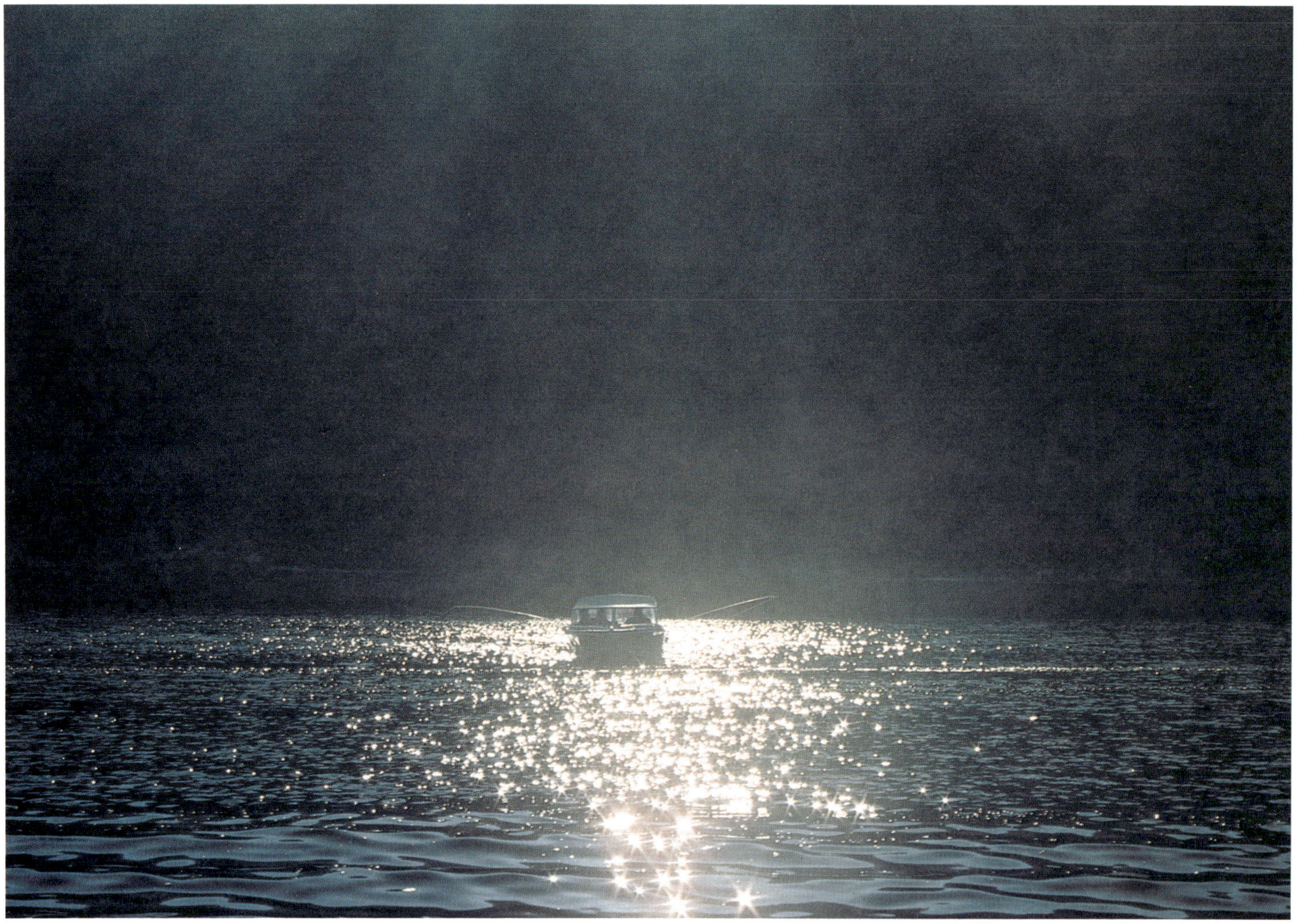

composition of the oceans with pollutants, most of which arrive by air. They include nuclear fallout, waste products of the burning of fossil fuels (including heavy metals and the oxides of carbon, sulfur, and nitrogen), PCBs, and insecticides.

Luckily for man, seawater has the ability to cleanse itself. Beyond the continental shelves lie 99 percent of the ocean waters; the visible damage to the marine ecosystem is in local inshore waters and estuaries. Since the population of British Columbia is concentrated in the southwestern corner of the province, most inlets and river mouth areas are free of sewage and garbage sludge. But elsewhere along the coastline of the Pacific Northwest, the seabed has been ruined by bark and wood fiber. In some places the rafting of logs and the dumping of sawdust have smothered entire shellfish beds.

Near the north end of Vancouver Island an open pit copper-molybdenum mine began dumping mine tailings into Rupert Inlet back in 1971. The company predicted the tailings would extend only 2 km into the inlet and remain 100 m (330 ft) below the surface. Six years later more than 70 million tonnes (77 million tons) of mine wastes had spread 15 km (9 mi) into Rupert Inlet and into nearby Quatsino Sound and adjacent Holberg Inlet. The unconfined discharge has obliterated bottom life and inflicted irreversible damage to the marine ecosystems. The ore deposit will be exhausted in another 20 years; the damage will continue long after that.

Mine wastes are a common source of heavy metal poisons, notably copper and zinc. Although fish require minute quantities of these substances to grow and reproduce, excess amounts can be lethal. Rainwater leaching through mine tailings and then draining into salmon streams has decimated entire stocks. Sublethal quantities cause salmonids to forego reproduction. So far as is known, salmon in the open reaches of the ocean are unaffected by heavy metal pollution. Although salmon are sensitive to the presence of copper and zinc in water, and protect themselves by avoiding contaminated areas, this may become more difficult for them as large and powerful mining companies learn how to exploit the deep seabed.

Manganese nodules—containing large quantities of manganese, cobalt, copper and nickel—are slowly formed on the floor of the ocean by deposits of minerals from seawater. The nodules range in size from golf balls to basketballs and are found in large areas of the Pacific and Indian oceans. With today's technology, the nodules, only partially buried in soft sediments, are fairly easily harvested. The use of bottom-crawling mining vehicles, supported by a submersible barge, is one method; others include compressed air suction devices and "endless bucket" systems for lifting the ore nuggets to the surface.

The environmental impacts of deep-sea mining are unknown. Mixing of cold deep water with warmer surface waters may affect ocean currents. Roiling of seafloor sediments may affect phytoplankton by reducing the light. Fish life will certainly be harmed by the discharge of tailings and the release of chemicals in whatever refining process may be undertaken.

A final and potentially more serious disturbance to the remote dark floor of the ocean is the proposal to bury atomic wastes at sea. Between 1946 and 1962 ocean dumping of liquid and solid nuclear wastes was permitted on both east and west coasts of the United States. Surveys made between 1974 and 1976 showed traces of plutonium and cesium contaminating the ocean floors near these dump sites. The United States government has banned ocean disposal of radioactive wastes pending further study.

European nations, however, continue dumping low-level wastes in the Atlantic. Scientists advising the European Nuclear Energy Agency feel that the possibility of radioactive materials affecting the marine food chains is slight. The fact is that they do not know what the effects may be. Certainly, each dumping is an irreversible step toward greater contamination.

The impossibility of calculating the extent of damage caused in the world's oceans by man-made pollutants is a major obstacle facing those who would preserve the ecology of the seas.

HOMECOMING HAZARDS

Pacific salmon and steelhead may spend from one to four years feeding and growing in the waters of the Gulf of Alaska and northern Pacific Ocean. Little is known of the causes of their mortality during these years at sea. Killer whales and seals prey on them in the open waters of the North Pacific, as do two primitive parasitic fishes: the lamprey and the hagfish. The toll exacted by these predators is not known.

In the year of sexual maturation the salmon begin their homing migration back toward the North American coast line. Waiting for them in the coastal waters and river estuaries are sea lions, harbor seals, killer whales, and the commercial and recreational fishing fleets. So efficient has the coastal fishery become that almost all the salmon entering a major river estuary could be caught by the fishery. Accordingly, limits are set on the catch.

Leaving the estuaries for fresh water, the salmon begin the last leg of their journey. As they start upstream they are preyed upon by seals, bears, gulls, and poachers. Of them all, man is the most efficient predator. Commercial fishermen may catch 80 percent of the returning adult salmon. Indians—with traditional rights to fish—take more of the survivors along the upstream routes.

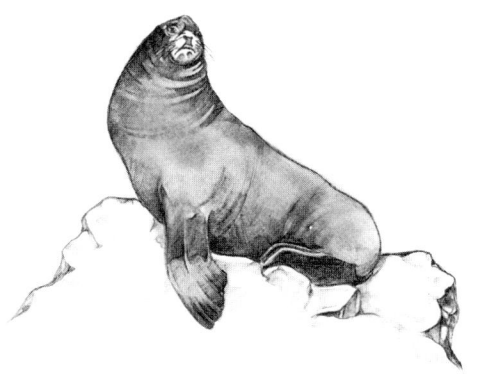

Sea lion

Grizzly bear

Poaching in Newfoundland and the Maritime provinces now threatens the existence of the Atlantic salmon. In Washington state, on the Pacific coast, white fishermen have turned to poaching to protest a U.S. District Court ruling of 1974 that Indian tribes are entitled to 50 percent of the salmon and steelhead catch in ancestral waters. In British Columbia in 1975, during a strike of fishermen, poachers forced the closure of the Fraser River fishing grounds when, on a single weekend, they took from a stretch below the Fraser Canyon tens of thousands of pink and sockeye salmon bound for their Chilcotin spawning grounds. Scarcely 1,000 spawners passed through the Hell's Gate fishways.

Low water levels that year created ideal conditions for poachers, who set nets at night across river pools where salmon rest. These interludes of quiet holding in pools are brief, for the upstream journey is long and time to complete it is limited. Once salmon enter fresh water they stop feeding and rely on body fats and protein for energy. The biological time clock is set. They must reach their spawning ground, deposit and fertilize their eggs before they die.

The ability of the individual salmon to withstand fatigue and exhaustion is the key factor in the success of the upstream climb. Like an athlete who is capable of lifting only so much, running so fast, or going so far, a salmon can only jump a certain height, swim at a certain speed, negotiate so many jumps or fastwater runs. Since they do not feed once they enter fresh water and so do not add to their energy bank, any changed condition on their migration route directly threatens the next generation.

A rock slide that partially blocks the river can convert what might have been just rapids into a roaring torrent. The water surges and boils, a cauldron of upwellings, eddies, and vortices. It may calm momentarily, only to roar back again. Merely to maintain position and direction at the foot of a high velocity chute drains the fish of energy. Robbed of its full swimming power the salmon can neither gain on the current nor leap over the obstacle.

Such a rock slide occurred on the Babine River in the Skeena watershed in 1951. Most of the Babine sockeye run of that year was blocked below the slide; those that managed to surmount the blockage were so weak and battered that few reached the spawning grounds. Between 30 and 40 percent of the dead female sockeye examined on the spawning area that year were found to have died unspawned. The famous Babine Lake sockeye run had been decimated.

This natural disaster took place 50 years after the dam fiascoes had ruined the sockeye runs to Quesnel-Horsefly and Upper Adams River. The Canadian Fisheries Department reacted at once to the news that numbers of dead and dying sockeye were floating in the Skeena River below Hazelton. An aerial search disclosed the fatal rock slide in a deep canyon above the confluence of the Babine and Skeena rivers. To clear the slide required a 100-km (60-mi) long access road and the construction of 15 major timber bridges and 40 minor ones. In the meantime, pack train work crews were successful in blasting partial pathways through the huge rocks that had tumbled to the foot of the slide. A third of a million fish were able to pass the barrier the following year, and by the next year the waterway had been restored to normal.

Natural slides are one hazard, but turbulent, unpassable stretches of water—also caused by unusually high water levels—are another. One year, conditions of late snowmelt and sudden runoff delayed the early Stuart Lake run of sockeye at Yale rapids on the Fraser River. The delay only lasted six days, but of the 30,000 to 35,000 spawners expected in the Stuart River that year, only 2,000 arrived.

Not far above the Yale rapids is Hell's Gate. This gorge is the narrowest portion of the 11-km (7-mi) long Fraser Canyon. It is located 52 km (32 mi) north of the town of Hope and 210 km (129 mi) from the mouth of the Fraser. Through it flows the water drained from 220 000 sq km (84,000 sq mi) of central British Columbia, an area comprising one quarter of the province. The sediment-clouded water froths and roils through this granite gorge, which is only 34 m (110 ft) wide at its narrowest point. Some stocks of salmon spawn 1000 km (600 mi) beyond here.

HELL'S GATE

For thousands of years migrating salmon fought to surmount the swirling cataract in this rock-walled strait, but at times of flood they could not. The average annual rise and fall of the water level at Hell's Gate is 18 m (60 ft). Since records were started in 1912, water levels of 28 m (90 ft) have been noted. In the great flood year of 1894, it is estimated to have reached 38 m (126 ft).

The commercial catch of Fraser River sockeye increased during the late decades of the last century and into the 1900s, culminating in 1913 in a record catch of 31 million sockeye. This catch has never been repeated. A man-induced catastrophe had occurred the winter before when rock from construction of the Canadian Northern Railway (later to become the Canadian National) was dumped into the river at Hell's Gate. It was a routine practice: rock had been dumped at other points along the Fraser Canyon as well.

In July the first fish of that 1913 record sockeye run appeared at Hell's Gate. Some of these managed to pass upstream in higher water but, as the summer wore on, alarmed observers reported that "incredible numbers" of salmon were congregated in all the eddies and creeks for 10 miles below the man-made block. Authorities recognized the problem but took until autumn to react. Some rock was blasted out that winter to provide passageways for the fish, but the stocks that migrate higher up the Fraser during July, August, and early September were largely destroyed.

The problem at Hell's Gate was made worse the following winter by a second slide of rock. The railway had driven a tunnel through a huge rock cliff on the east bank of the river. This cliff collapsed and 75 000 cu m (100,000 cu yds) of rocks tumbled into the river. In the low water periods of that year frantic efforts were made to restore the river channel to its natural condition.

By March 1915 some 48 000 cu m (60,000 cu yds) of rock had been removed, but the river bed never returned to normal. The rock deposited in 1913 had produced a drop of 1.5 m (5 ft); the rock slide of 1914 increased it to 4.5 m (15 ft). The best the engineers were able to do was reduce the drop in water level through the gate from 4.5 m to 2.75 m (15 to 9 feet).

In the decades that followed, the commercial catch of Fraser River salmon steadily declined. Investigations showed that the block at Hell's Gate was the major cause. Although spawnings below the Gate were not affected, many of the larger upriver stocks had all but disappeared. To rectify the situation two major fishways—one on each bank—were built between 1944 and 1946, and they are responsible for the partial restoration of many of the upriver Fraser runs. But without the efforts of local fisheries officers before World War II to help salmon past the Hell's Gate block, there would have been insufficient numbers of the different stocks of salmon left to rehabilitate.

Hazards

Log booms in Nanaimo harbor area

Clear-cutting, interior British Columbia

Log bundling, Babine Lake

Open-pit copper mine on island in Babine Lake

New plant growth in river estuary

Natural erosion of estuary banks

Nanaimo River estuary

Bridge and railway construction can adversely affect salmon streams

Fraser River delta (Steveston fishing fleet in foreground). Man's activities encroach on British Columbia's major salmon river estuary

Natural erosion causes stream blockage

Highway construction with engineered banks to protect salmon stream

Pulp mill (background) and gravel pit pose different hazards to salmon streams

Forest destroyed by fire can result in erosion until second-growth restores soil stability

Sea lions in Georgia Strait

Migrating chinook fighting rapids at Stamp Falls . . .
. . . instead of taking easy way up the fishway

Enhancement

The history of the earth is of species appearing and disappearing; of new species evolving, then receding. Each animal—save man—has a habitat, a place where it is generally found and to which it is adapted. Several or many species may share a habitat, but each has its own special locale, or niche. No two species occupy the same niche.

If habitats change, niches must also change. If an animal cannot adapt to the changes, it becomes extinct. Natural cataclysms occur and take their toll. But successfully adapted species can survive these hazards. Earthquake, flood, drought, or loss of basic food supply seldom imperil an entire species. Nature compensates. Good years follow bad ones; sometimes several lean years are followed by a period of plenty.

One of the biggest disrupters of nature's balance appears to be ourselves. Man the superpredator. Even prehistoric men may have hunted species of animals to extinction. Their descendants, the Plains Indians, used fire-drive methods to stampede entire herds of buffalo over cliffs to their destruction. In the past, man destroyed habitats through agriculture and deforestation. Today's technological man accelerates the carnage.

When Europeans first arrived on the north Pacific coast, an ideal habitat existed for salmon. Rainfall was heavy; the high coastal mountains collected winter snow, and the forested slopes and valleys conserved the moisture, feeding it gradually to the streams. Thus a supply of clear cold water was assured to attract the adult salmon and to nurse the eggs and fry.

Man the predator at that time was Indian. Tribes dwelling along the upper reaches of the Fraser, Skeena, and Nass rivers were totally dependent on the annual crop of salmon. They took thousands each year, but never so many as to weaken future runs. They harvested the fish at whitewater points of constricted waterflow where they camped yearly. To catch the salmon, they used dip nets, spears, hooks and lines, haul seines, weirs, traps, and jump baskets. These baskets were set near waterfalls to catch salmon falling back.

Said the 1906 annual report of the Canadian Department of Marine and Fisheries about one weir across the Babine River: "There were posts driven into the bed of the river which is 200 feet wide and from two to four feet deep and running swiftly . . . Then sloping braces well bedded in the bottom and fastened to the tops of the posts, then strong stringers all the way on top and bottom, in front of the posts, then a panel beautifully made of slats woven together with bark set in front of it all . . . This made a magnificent fence which not a single fish could get through." But obviously some spawners got through; otherwise the Babine salmon run would have ceased to exist.

The white man, of course, had noted the plenitude of salmon and recognized its value. Still, the distance from European and other markets was overwhelming. There was no way of preserving the product. This changed with the advent of the canneries. The first commercial canning of Pacific salmon took place on the Sacramento River, California, in 1864. Two years later the company transferred its operation north to the Columbia River, Washington. The first cannery on the Fraser opened for business in 1870, producing 30,000 cans of salmon. By 1878 the Fraser River business had expanded to ten canneries and a pack of 5.5 million cans.

In the season of 1885, it is said, "the Fraser was choked with sockeye." This was not uncommon. Annual catches of Fraser River sockeye in the dominant cycle years went up to 30 million fish. But this did not last. The early years of commercial fishing were marked by carelessness and greed; entire runs vanished, destroyed by overfishing or loss of habitat.

In the 1920s the total British Columbia catch of sockeye, pink, chum, chinook, and coho was 186 million pounds per year. In the decade of the 1930s it fell to 164 million; in the 1940s, 155 million; in the 1950s, 137 million. This was an all-time low. The decline coincided with the industrial boom that followed World War II, which brought further pressure on the already depressed stocks. Overfishing, illegal fishing, and lack of management knowledge were longer-term causes of the problem.

In the late 1950s the Canadian government moved to stop the decline. The Fisheries Department increased its conservation, enhancement, and research staffs. The efforts of these men were at least partially successful: by the end of the 1960s the ten-year average catch had risen to 139 million pounds. This figure stabilized during the first half of the 1970s, yet the balance remains precarious. Most gains from improved management over the past 20 years have been nullified by losses through environmental damage and by the capture of Canada's Pacific salmon during their oceanic migrations.

On the assumption that the 200-mile limit and international agreements will protect the salmon from being overfished by foreign fleets, the Canadian government has launched a program to restore west coast salmonid stocks to their pre-1900 abundance. The plan is based on the

High water flows–autumn

Dry river bed–summer

results of more than 25 years of development of spawning channels, flow control systems, hatcheries, and rearing facilities.

Bottlenecks in salmon production are to be found in freshwater and estuarine areas; accordingly, initial emphasis is on improving freshwater habitat. Stream rehabilitation, obstruction clearance, and reforestation are being undertaken where needed.

The simplest steps to salmonid enhancement are those which regulate the supply of water. Whenever a stream that is low in summer can be fed more water to increase the flow, fish living there will benefit. Wherever the dangers of violent floods can be reduced, more salmon eggs will survive.

Equally simple are the techniques of assisting adult salmon in their migration. The removal of a log jam or the provision of a small fishway will help the spawning salmon get to the places where their young can thrive.

Another aspect of the Salmonid Enhancement Program involves man-made facilities. Spawning channels, hatcheries, and fishways are principal among these; incubation units, rearing ponds, stream engineering, flow and temperature controls on spawning grounds are associated works.

Salmon are sensitive creatures, finely tuned to the life they lead. Canadian Fisheries Department planners are well aware that efforts to help the salmon, regardless of good intentions, may be harmful rather than helpful. The "soft life" in a hatchery may only predispose young salmon to an early death when they face natural conditions. The artificial incubation of eggs may produce fry less vigorous than their natural counterparts. These and other lessons of salmon culture have been learned in a hundred years of research and trials from which two principles have evolved: first, the less the natural history is changed, the less the risk of harmful effects; second, salmonid enhancement is far more difficult than it seems.

For example, it seems simple to open up a new watershed for salmon to colonize. Dozens of streams support salmon only in their lower reaches, and a fishway past a falls would seem to be the answer. Some streams may have had salmon in them years ago but are now barren. Why not put some salmon into these streams? Unfortunately, all salmon are not colonizers, and of those that are, not all are successful. It takes luck to develop a new run. You need the "right" salmon and you need to get the "breaks" to build a population large enough to persist if conditions become unfavorable.

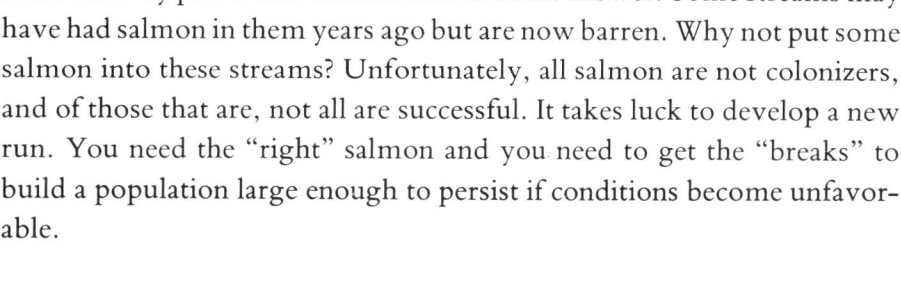

Courting chum pair

GENETIC RESEARCH

In contrast to other salmonids, pink always mature at two years of age; their life cycle is fixed. This means that two entirely separate pink salmon populations exist which never interbreed: one matures in odd-numbered years and the other in even-numbered years. In some watersheds one of the cycles is missing; it is said to be an "off" year for pink (the following fishing season being an "on" year). The major North American areas showing this pronounced on/off character are the Fraser River and Puget Sound streams, and some streams in the Queen Charlotte Islands. They have been this way since before the white man; the condition probably existed at the time of the last Ice Age, 10,000 to 12,000 years ago.

Why pink salmon fail to populate certain river systems one year but do so the next is unknown. Without doubt, part of the reason is the inflexible two-year life span which prevents the overlap of generations that occurs in other salmonids.

Off-year cycles of pink salmon represent a biological opportunity for scientists. The Fraser River, for example, produces millions of pink in odd-numbered years and none whatever in even-numbered years. Since the early 1900s fish culturists have seized this opportunity to experiment by transplanting pink runs into the Fraser's even-year "off" cycle.

Assisting pink salmon transfers by genetic techniques began in 1969 when Canadian researchers showed that injection of salmon pituitary gland hormones resulted in the males—but only the males—becoming mature one year earlier than normal. This discovery suggested that it might be possible to transfer one-half of the heredity of a successful on-year run into the barren off year by fertilizing the eggs from a donor stream with milt of "accelerated" males from an on-year run. Subsequent tests revealed that the hormone-injected juvenile males were in fact fertile and that under test conditions their milt was as potent as semen from wild fish.

At the same time another research team working closely with the hormone-injection group developed methods for deep-freezing salmon sperm, much in the manner that bull semen is preserved for artificial insemination of cattle. Building on preliminary work done in the United States, Canadian researchers have brought the flash-freezing technique to the point of practical use. Sperm from a stream's on-year run can be stored and used to fertilize several million donor eggs for planting into the same stream's off-year cycle. Like those from the accelerated males, these

Cross-section of spawning channel

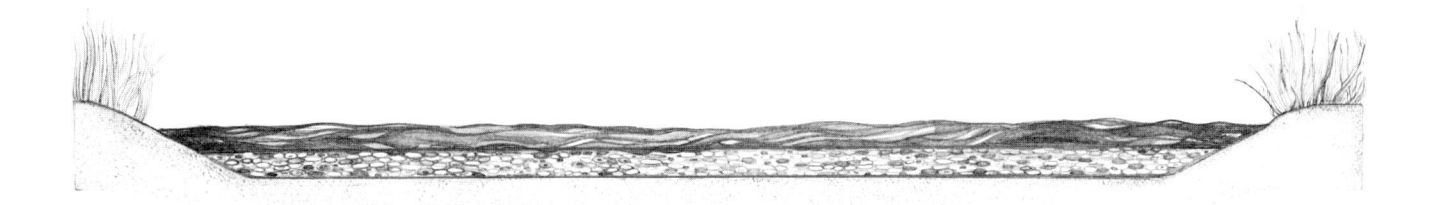

offspring will carry one-half the heredity of the successful on-year run.

The aim of yet another research group is to postpone the maturation of pink for a year. The strategy involves deceiving the fish into maturing over three years by using artificial light to vary the apparent length of day. To the fish, whose growth is being retarded by water temperatures and diet, only two years seem to be passing.

SPAWNING CHANNELS

Less than one percent of salmon eggs survive to become adult salmon, and for all species the greatest losses occur in the egg and larval stages. The survival of salmon eggs is limited by low or high stream flows and by the silting of the spawning beds that restricts the flow of water through the gravel. If these factors are controlled, and optimum stream conditions prevail, a greater production of young salmon results. Artificial spawning channels evolved from scientific efforts to increase survival rates by improving salmon breeding grounds.

By providing favorable water depths, controlled flow, and good gravel porosity, a spawning channel reproduces the best conditions of a natural salmon stream while avoiding the total artificiality of a hatchery. Fry from a spawning channel emerge from the gravel at the same time as the fry in the nearby stream. They are thus more properly timed to the environment and comparable in physical condition to wild fry.

The first opportunity to test the spawning channel concept in British Columbia came when the B.C. Electric Company announced plans for a power project on Jones Creek, near Hope, in the early 1950s. This small stream is the home of pink salmon. The proposed dam, by drastically reducing stream flow at critical stages of the salmon's life, would have eliminated the run. The company agreed to build a spawning channel to compensate for changes in the natural system.

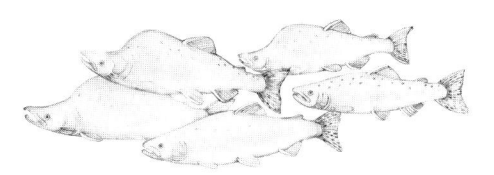

School of mature pink prior to spawning

Pink salmon first spawned in this artificial channel in 1955. The eggs remained in the gravel through the winter, incubating under a flow of fresh cool water. The following spring, pink salmon fry were trapped and counted as they moved into the Fraser River on their way to the ocean. Thirty-eight percent of the eggs were found to have survived as seaward migrating fry, a fourfold increase over survival in the wild.

Simultaneously, a different approach was taken by the International Pacific Salmon Fisheries Commission at Horsefly Lake in the Quesnel area. Ponds were constructed with gravel laid over perforated pipes. Water, pumped from the lake into the pipes, upwelled through the gravel at controlled rates. Sockeye salmon eggs—some planted and some naturally spawned in these ponds—showed high survival rates, ranging between 20 and 68 percent. This technique works best with sockeye stocks that spawn in the still water of lakes rather than in fast flowing streams.

Spawning Channels

The success of the Jones Creek and Horsefly installations confirmed not only that salmon will spawn in such manipulated conditions but also that these conditions will result in more fry surviving than do in nature.

During the late 1950s and early 1960s, government agencies in British Columbia, Washington, and Oregon built several spawning channels in attempts to offset salmon losses at power dams. Many of these spawning channels failed because their locations were dictated by the site of the dam that supplied the channel with water from its forebay and with fish diverted from its fishway. The spawning channels then were often well downstream of the areas where the parents of most of the salmon had spawned and where the fish had begun their lives. As a result, the channels were stocked with mixtures of fish that were not adapted to the new conditions created by the downstream location of the channel.

The consequences of such a defect were dramatically shown with the experience of coho salmon in the Robertson Creek research channel, which was built on a secondary outlet of Great Central Lake in the Stamp River system on Vancouver Island. Coho that entered this channel were fish of mixed origins, some that normally spawned in the warm waters of the Stamp River and others that moved through Great Central Lake to spawn in the cool tributary streams of the upper reaches of the lake. Not only was this second group of fish afflicted by the unaccustomed high water temperatures of the channel but also their biological urge to move farther upstream was thwarted. They died unspawned.

In the United States, the same situation arose with chinook salmon in a channel built to compensate for losses suffered at the McNary Dam on the Columbia River: salmon that were destined to spawn many miles upstream in the cool waters of the Snake River died unspawned in the warm waters of the channel at the downstream dam site.

The idea of using spawning channels to improve production of salmon, rather than to offset the losses caused by power dams, originated in Canada. These federally funded channels were built in locations adjacent to natural spawning areas. Good results were achieved every time.

The first of these production channels was built in 1961 by the International Pacific Salmon Fisheries Commission at Seton Creek in the Fraser River system near Lillooet. The channel was designed to augment pink salmon populations that had been adversely affected by a hydroelectric power dam built on Seton Creek five years earlier. A second channel at this site was added in 1967. The Commission followed through in the Fraser system with successful spawning channels for sockeye at Weaver Creek, near Harrison Lake, and at Gates Creek. The most recent is the Nadina channel, on the headwaters of Francis Lake in the Nechako system, which came into production in 1973.

Fulton River spawning channel complex. Babine Lake in background. FRANK VELSEN

Spawners entering Fulton River are counted

Babine River counting fence

Fisheries biologist selecting fish for controlled experiment in sockeye spawning

Fish are sorted by age and size for experiment

Hooked upper jaw is characteristic of all Pacific salmon spawning males

Technician carrying selected salmon to pens

Aerial view of fish pens with observation towers in Fulton River spawning channel

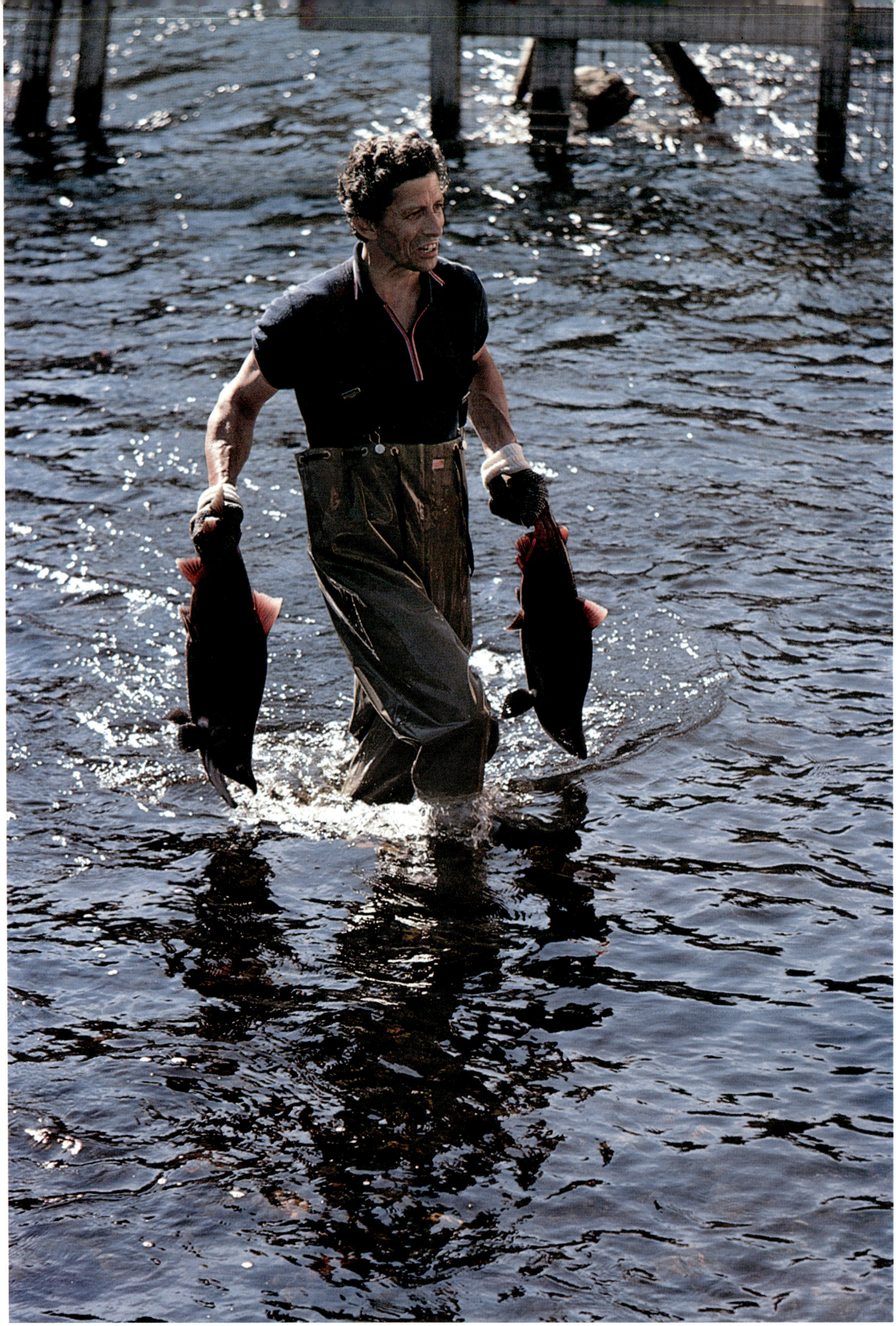

Fish are marked for observation with numbered Peterson disc tags.

Scale sampling for age identification of chum at Qualicum River spawning channel

Otolith (inner ear bones) are also used for age determination

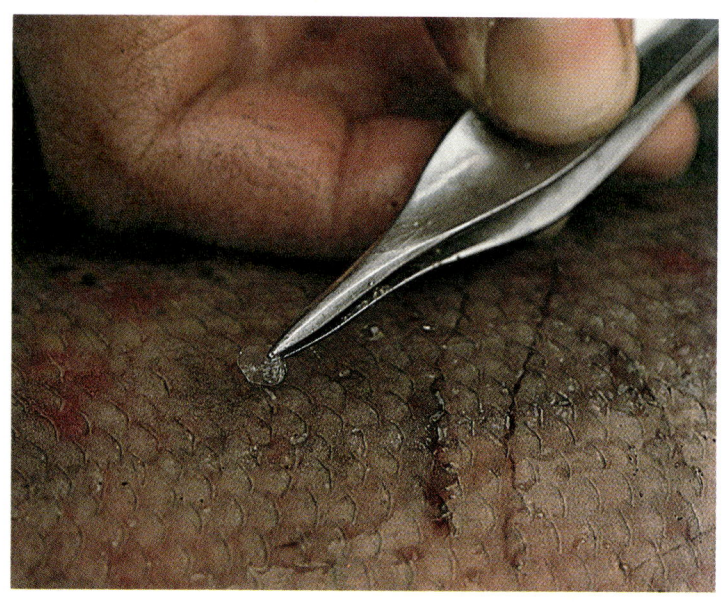

Fish scale

Transfer of salmon at Fulton River. KEES GROOT

Smolt trap at outlet of Babine Lake

The success of the first production spawning channels in the Fraser River watershed was encouraging. Nonetheless, the Fraser was suffering from 40 years of overfishing and, by the 1950s, from the acute effects of "civilization." The magnificent Fraser salmon runs were under attack and declining. Canadian Fisheries scientists were looking for a means to offset the losses. Since one of the problems was environmental destruction, they looked elsewhere for pristine conditions.

The Skeena River watershed in north-central British Columbia was still in its natural state and contributing half a million sockeye annually to the commercial fishery. Yet studies indicated that the system's most important nursery area, Babine Lake, was underutilized by yearling sockeye. Further research revealed the cause to be lack of tributary spawning areas; that is, insufficient spawning streams to enable more fry to survive and move into the lake. Artificial spawning channels were constructed near Babine Lake to increase sockeye fry entering the lake and so attain maximum sockeye production from the Skeena River system.

It is the Canadian government's most ambitious spawning channel project to date. At a cost of $8.5 million, the lake's two major tributaries, Fulton River and Pinkut Creek, were extended by the construction of artificial spawning channels, plus a combination of dams, tunnels, and pipelines to provide controlled waterflow to the channels and the two natural streams. The complex was designed to increase annual sockeye production by 125 million fry. At a one percent fry-to-adult survival rate, the project has the potential to produce an additional 1,250,000 adults, of which one million will be harvested by commercial fishermen at the mouth of the Skeena River.

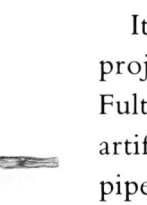

Trilevel intake to control water temperature

INCUBATION UNITS

Where a spawning channel is needed but cannot be built, or is too expensive to build, an incubation unit may be the answer. Salmon eggs, fertilized by hand as in hatchery production, are placed in gravel contained in large plastic, wooden, or concrete boxes. The eggs incubate in a controlled flow of fresh cold water directed upward through the gravel.

The concept was pioneered on the West Coast in the first two decades of the present century. These early efforts were frustrated by silt problems in the water, but work by Russian scientists in the 1920s kept the idea alive. Continuing research at the Pacific Biological Station, Nanaimo, B.C., culminated in the 1960s with incubators producing eight times as many pink salmon from the same number of eggs as were produced from adjacent natural spawning areas.

Five different types of gravel-box incubators are presently in use along the Pacific coast of North America; all require a well-oxygenated, plentiful water supply containing little or no sediment. The boxes employ features of both spawning channels and hatcheries but are much smaller in scale. Once the mixture of eggs and gravel has been placed in the box, incubation, hatching, and development to the fry stage occur in natural sequence. The fry are allowed to escape with the outflow water to the nearby stream, or are collected for counting (and/or fin clipping) before being transferred to either rearing ponds or the nearby watercourse.

Two recent developments include a shallow-gravel box design and plastic artificial turf in place of gravel. The replacement of gravel with plastic materials has been under evaluation since 1969. In addition to the artificial turf, specially designed plastic pieces have been tested. A recurring problem with the plastic substitutes has been the premature emergence of the fry, sometimes weeks early. Whether it is due to the shape of the replacements or to chemicals leaching from the plastics, no one knows yet.

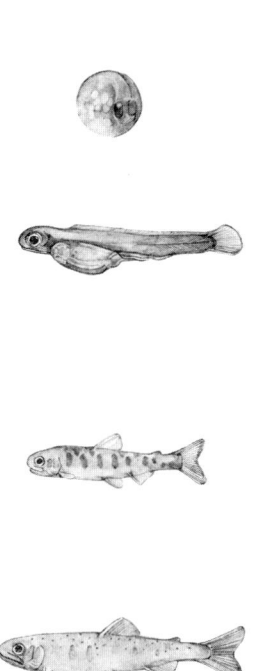

More research is underway into such things as the relative merits of crushed or round gravels, egg-loading densities, fry quality, and survival. But these are of secondary importance; the knowledge of the many subtle factors that spell success or the lack of it is already with us. Fish culturists know the requirements of water temperature and oxygen concentrations, the need to shield developing alevins from the light, the potentials for epidemic disease, the conditions when the fry hatch, and the changes in those conditions that trigger migration. Given an understanding of these essentials, the incubation box is a good tool for salmonid enhancement.

Sockeye egg, alevin, fry, smolt

HATCHERIES

Canada's renewed interest in hatcheries began in the mid-1960s. A small experimental hatchery was built on Big Qualicum River, Vancouver Island, in 1967. Encouraged by the results, and with the prospect of contributing significant numbers of salmon and steelhead trout to sport and commercial fisheries, a major hatchery program was begun in British Columbia. The first hatchery in this series was built on the Capilano River, North Vancouver, in 1971. Since then the hatchery on the Big Qualicum River has undergone a major expansion; new hatcheries have been built on the Quinsam River and on Robertson Creek near the outlet of Great Central Lake, and a major new hatchery on the Puntledge River will be completed in 1980. These are all located on Vancouver Island. Construction is also in progress on several Japanese-style chum hatcheries in which warm ground water accelerates incubation so that the eggs hatch as much as six weeks early. By feeding the fry, it is possible to release them at the same time as natural fish but at a much larger size.

The most suitable species for conventional hatchery production are steelhead, coho, and chinook. The life histories of these species in the hatchery are similar to those of wild populations.

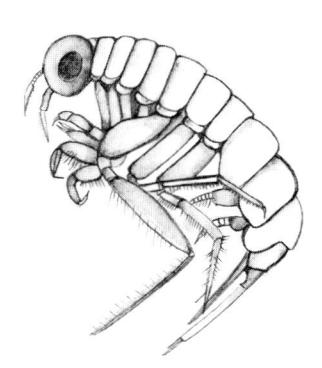

Hypiriid amphipod (3 mm)

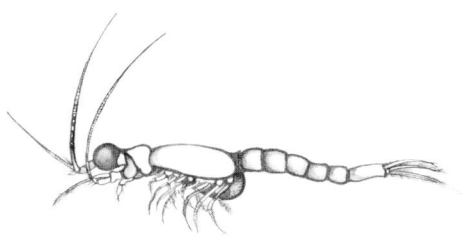

Neomycis rayii (3 cm)

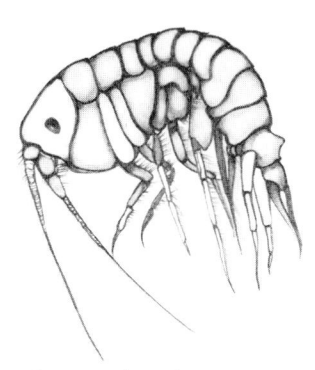

Gammarid amphipod (1 mm)

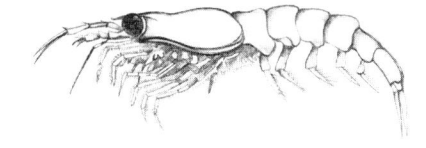

Euphasiid (3 mm)

On hatching in the wild, chinook fry normally spend three to four months in the stream or estuary of a major river before migrating to sea. (A few populations, known as "stream type" chinook, remain up to 15 months in fresh water before migrating.) Once at sea the juveniles feed on plankton, crab and shrimp larvae. As they grow they prey on larger crustaceans such as euphausids and mysids. Subsequently, they change to a fish diet consisting of herring, needlefish, and the like.

After 15 or 16 months at sea some of the jacks will return to their home stream to spawn. A larger number of males will return as three-year-olds, whereas the majority of females return as four-year-olds. Lesser numbers of both males and females return at five years of age and occasionally at six.

Chinook normally spawn in October, the upstream migration taking place during August and September. In British Columbia, chinook are primarily found in the larger river systems. They have been known to exceed 45 kg (100 lbs) in weight.

The life cycle of hatchery-bred chinook is similar to that of their wild cousins; however, conditions can be manipulated in favor of the young fish. By maintaining winter incubation temperatures in the hatchery at 9° C and rearing the fry to smolt size at 10 to 11° C, it is possible to release these fish at a size of 5 grams (90 smolts per lb) after approximately 100 days of rearing. Average rate of survival-to-adult for chinook smolts released at this size before July 1st is about 3 percent.

Higher survival rates can be achieved by holding chinook smolts for a full year's rearing and release at a size of about 90 grams (5 per lb). The full implications of extended rearing have not yet been determined, but a number of such experiments are in progress.

Coho salmon are smaller than chinook. After emerging from the gravel in March or April the fry remain as stream residents for about 14 months before migrating to the sea. Once at sea they grow rapidly on a diet of zooplankton and, to a lesser extent, fish. After four or five months at sea some jacks return to the home stream. These jacks, like the normal adult coho, die after spawning.

The majority of coho stay at sea for 16 to 18 months, returning to spawn as three-year-olds. Some coho stocks originate in glacier-fed cold-water systems. These may spend two years in fresh water before reaching migrant size and will be four years old at maturity. Coho normally spawn in November, but adult migrations into the home stream may start as early as June and continue until January, with some late spawning taking place in February.

Hatchery reared coho juveniles grow at the same rate as their counterparts in the wild. The rearing extends over a 15- to 16-month period after

Hatcheries

which the smolts are released at a size of 30 grams (15 per lb). Between 5 and 30 percent survive to become adults.

Hatcheries supplied with warm spring water, or having an inexpensive source of heat such as discharge waters from thermal plants, could accelerate the production of adult coho by producing smolts of migrant size in the first summer of rearing. The adults would be two years old rather than the normal three. This type of control has been demonstrated experimentally, but few hatcheries have a warm water supply and the cost of fossil fuels prohibits heating large quantities of water for fish rearing.

In all salmon hatcheries in British Columbia some facilities are devoted to the production of steelhead trout. Steelhead are not actively sought by the commercial fishery, but they are prized for their fighting qualities by anglers, and most are taken in their home streams. The life history of the steelhead is similar to that of coho.

In the wild, steelhead fry emerge from the gravel to take up residence in the stream for one to two years before migrating to the sea. They return to the river in their third, fourth, or fifth year of life. Although steelhead may enter their home stream in any month of the year, they generally spawn in April or May, sometimes after being nearly a year in fresh water without feeding. Like the Atlantic salmon, steelhead trout do not invariably die after spawning; some survive to spend another year at sea, returning to spawn twice or even three times.

Hatcheries attempt to produce steelhead smolts weighing at least 90 grams (5 per lb) or larger in one year, but this is difficult to achieve unless heated water is available. For this reason most hatcheries release steelhead at a size of 18 to 45 grams (10 to 25 per lb), which is less than optimal; the alternative is to rear the juveniles for a second year. If they reach 90 grams after one or two years of growth, about five percent will return to the river.

TRANSPLANTS

Replacing spawning gravel with hatchery incubators and adding warmer water for initial rearing can overcome many of the problems found in nature and permit successful transplants. For example, the Capilano River in North Vancouver never had a chinook run. Before a hatchery was built on this river, chinook smolts from the Big Qualicum River stock were brought over and released in the Capilano. These smolt releases were made in the summers of 1969-71 in batches of thousands, ranging from 52,000 to 112,000. Fewer than 100 adult chinook returned from these smolt releases, a not surprising fact given the difficulties experienced with chinook transplants.

In 1971, the first year of operation of the Capilano hatchery, chinook eggs were brought from Big Qualicum. More than 338,000 juveniles were

Capilano salmon hatchery

Hatchery technician

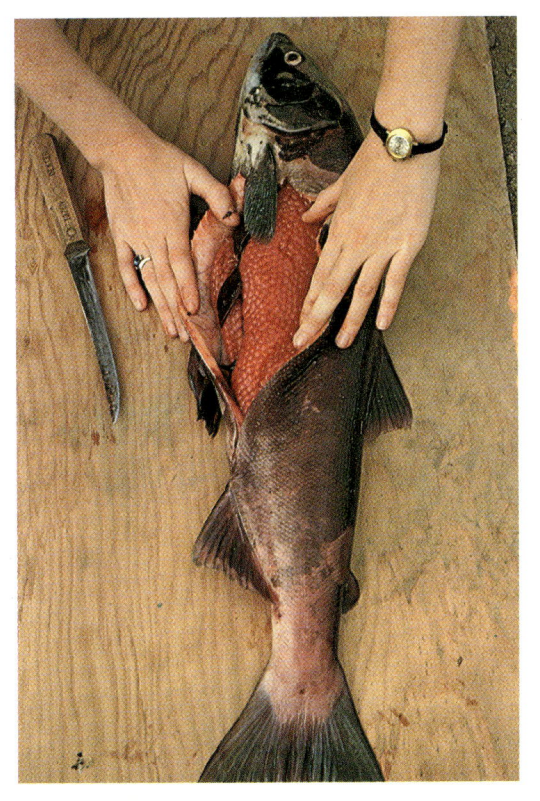

Female sockeye average 3,000 eggs

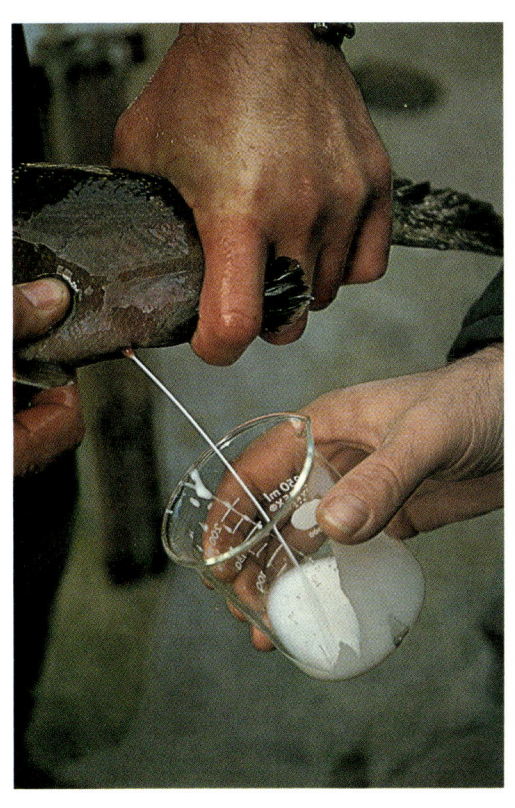

Milt is squeezed from male for artificial spawning

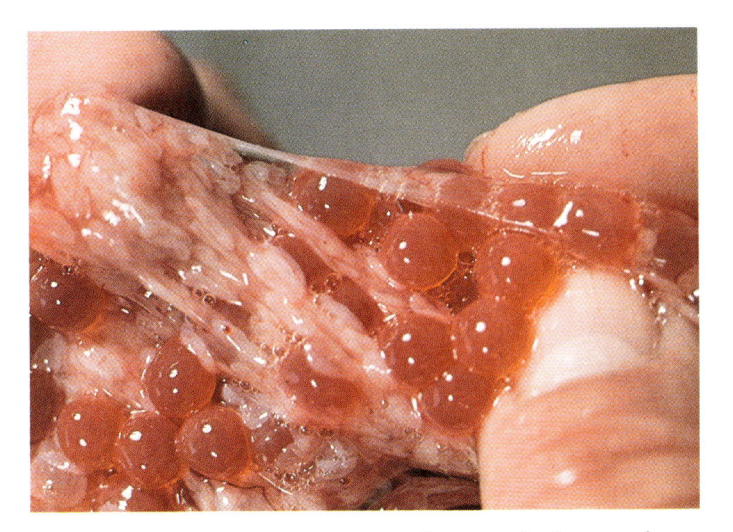

Eggs attached to membrane

Milt and eggs are mixed for fertilization

Four-cell stage: four cells are buoyed up by lipids attached to the yolk

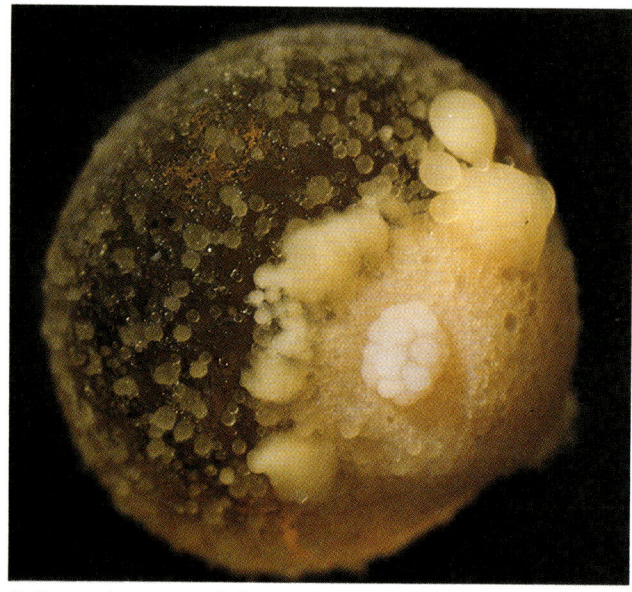

Cells continue growth by division, each cell dividing into halves. Shell has been entirely removed

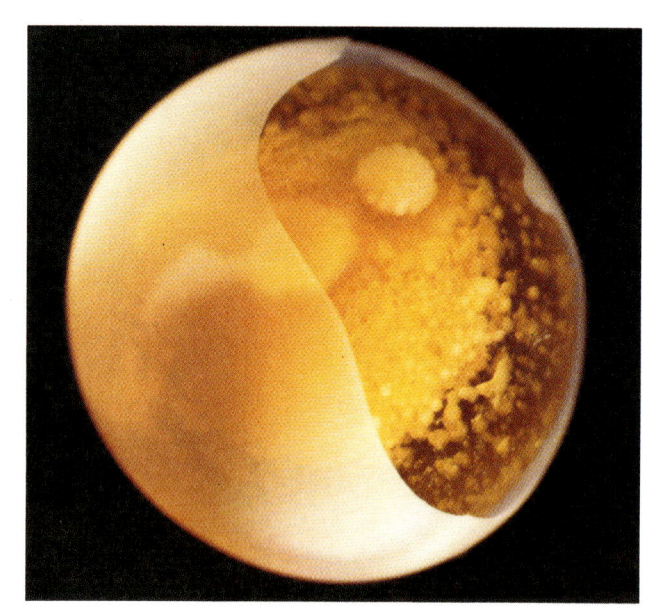

64-cell stage

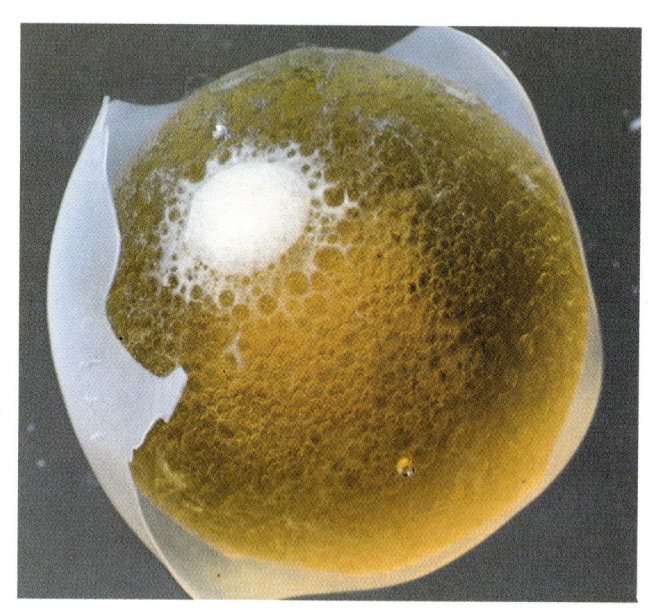

Cells have multiplied to the point where individual cells can no longer be distinguished

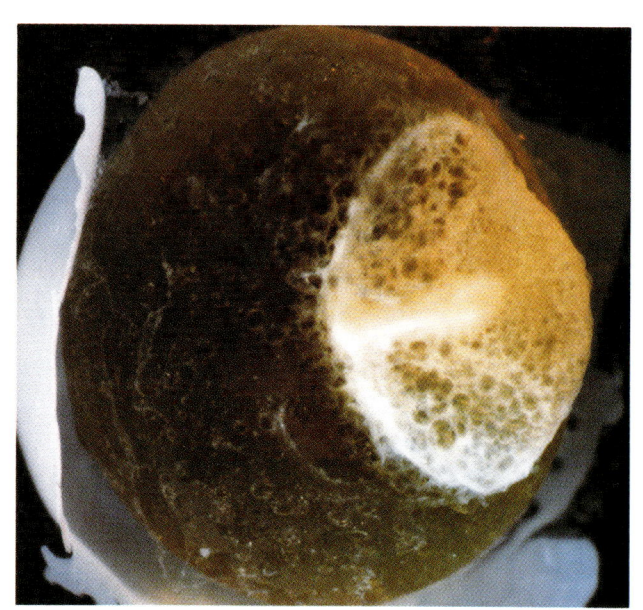

Cells are now beginning to overgrow entire yolk; early embryo can be seen

Overgrowth of yolk is nearly completed; embryo can now be easily seen

STAGES OF EGG DEVELOPMENT

Eyes are fully pigmented; embryo is ready to hatch. Cardiac vein can be seen at tip of nose
Photos: Frank Velsen

Coho alevin with yolk sac

Coho fry

Coho smolt

Gravel incubation box at Bear Creek for salmon eggs. Emerging fry collect in pails

Technicians counting fry from incubator

Pink fry raised in incubation box

Rosewall Creek experimental hatchery

Coho juveniles held for stock identification studies

Chum fry in rearing ponds are fed by automatic feeders

Fin clipping of chum fry at Rosewall experimental hatchery

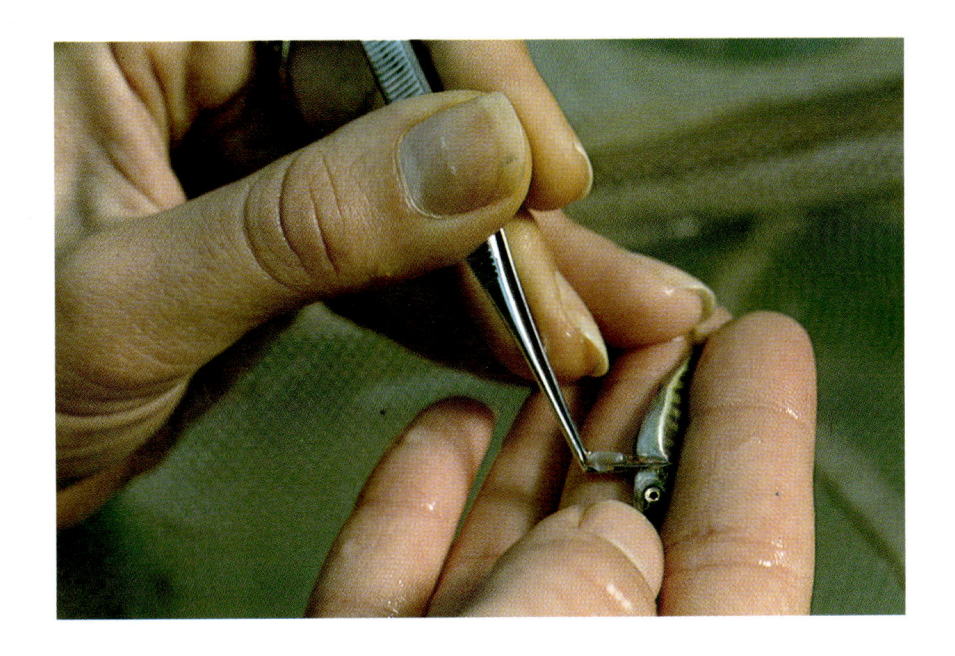

Experimental study on best gravel composition for salmon egg incubation

Pilot fish farm in Departure Bay, Nanaimo. Pacific Biological Station in background

Salmon reared in fish farm

Fish farming potential demonstrated by size difference of wild and artificially raised salmon of same age reared and released into the Capilano. This resulted in a return of over one hundred two-year-olds and nearly as many three-year-olds. In the second cycle of chinook rearing, half of the eggs brought from Big Qualicum were fertilized with males that had returned to Capilano from the previous smolt transfers. The number of two-year-olds returning to the hatchery from this release went to 700 in 1974. By 1977 the chinook return had increased to 1,167 fish. It is now apparent that this transposed run has become self-sustaining and does not require further infusions of eggs or smolts from Big Qualicum.

The success of this transplant is dependent on the continued operation of the hatchery. Suitable gravel areas for spawning are not found in the Capilano River itself, and the low water temperatures in the river prevent wild chinook smolts from growing large enough to migrate in their first summer.

HATCHERY OPERATIONS

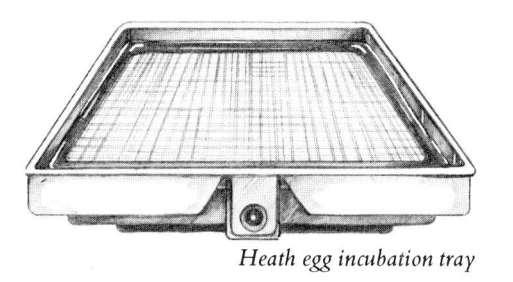

Heath egg incubation tray

Heath tray egg incubator

Most hatcheries are built with ponds suitable for holding adult fish. Depending on the particular run, the adults enter the hatchery via a fish ladder or are caught and trucked into the site a few days, weeks, or even months prior to spawning. As they reach final maturation the fish are checked every few days for ripeness. When the females are judged to be ready—the eggs being loose in the body cavity—they are anesthetized, then killed by a blow on the head, and the eggs are removed from a slit made in the body wall. The males are not killed; after the milt has been extruded by hand pressure along its sides, the fish is returned to the holding tank. Each male salmon can be anesthetized and "stripped" for its milt four or five times.

After the eggs from four or five females have been collected, they are mixed with milt and stirred. The fertilized eggs are poured into screened fiberglass trays, and then placed in a vertical stack incubator. The trays, holding 5,000 to 7,000 eggs, are held in stacks ranging from 8 to 20 trays high.

Clean water is run into the top tray. The water percolates through the screening and eggs of the first tray and then spills over into the tray beneath, and so on until it discharges from the tray at the bottom. The water carries with it a fresh supply of oxygen, without which the eggs or fry would perish, and carries away the by-products of metabolism. Vertical stack incubators are arranged in rows, usually in a room within the hatchery building which has reduced lighting.

The time required for the incubation of salmon and trout eggs depends on the temperature of the water. It takes approximately 50 days to hatching if the temperature is 10° C; longer at colder temperatures and shorter at higher temperatures. Once the alevins have hatched they are dependent for the next five or six weeks on the contents of their yolk sacs. When they become more fish-like and begin to search for food, they are ready to be transferred to the rearing ponds.

At this time the pond water temperature should be in the range of 10 to 11° C. A temperature of 5.6°C is about the temperature when chinook and coho fry begin to feed, but only at temperatures over 10° C does rapid growth occur. To obtain warmer water some hatcheries use groundwater from wells or controlled flows from the warmer surface waters of lakes; sometimes they heat the water to the desired temperature.

The fry are about 2.5 cm long at the time of ponding and begin feeding better if they are kept in fairly crowded conditions. As they grow they are redistributed to other ponds to prevent overcrowding. This is a critical stage, for if the fish do not learn how to feed, they waste away and die. The food used at this time is a finely ground mash which floats on the surface.

A common diet used in chinook and coho hatcheries combines a pasteurized wet fish mixture and a dry meal laced with vitamins, which comes as manufactured pellets of various sizes, packaged and frozen. The dry meal mix may consist of cottonseed, soya bean, wheat germ, herring, or anchovy; the wet mix incorporates pasteurized salmon or tuna viscera and herring.

The ponds used for fish rearing are of several types. The first to be used was the earthen pond, a simple excavation filled with water and having an inlet and drain. Earthen ponds are still in use. They may be large and circular or long narrow channels. Modern ones frequently have plastic or asphalt liners to permit cleaning and to prevent water loss through soil porosity.

The next stage in pond development was the concrete raceway. The water enters one end of the raceway, passes down its length, and discharges at the other end. Although raceways have given way to more complex pond designs there is a renewed interest in the raceway because of lower construction costs and ease of mechanization.

In circular ponds the water comes in under pressure through small nozzles near the periphery, giving the water a circular motion. The discharge water is removed via a center drain. Circular ponds of more than eight meters in diameter have been found unsuitable for salmon rearing, but smaller ones of fiberglass are now used as starter ponds for fry. Once the fish have begun to swim and feed they are transferred to conventional ponds.

The most recent and widely used design is the Burrow's pond, which is rectangular in shape and has an incomplete center wall. The water is introduced by two jet headers at diagonally opposite corners. This arrangement causes the water to circulate, and it discharges from the pond

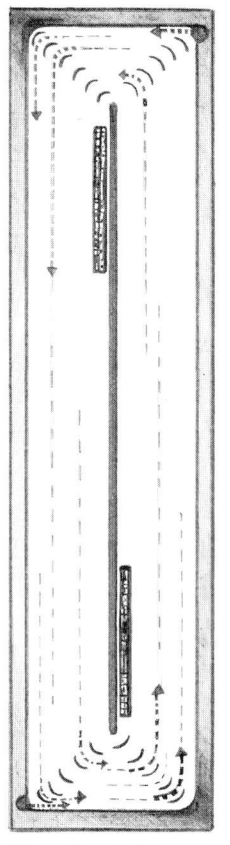

Water flow in Burrow's pond

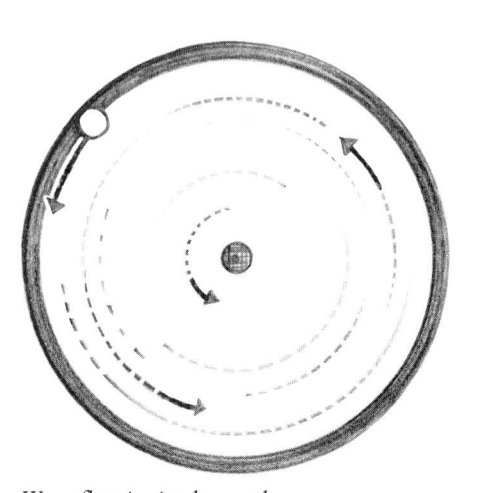

Water flow in circular pond

through two floor drains located in the lee of the center wall. Waste material in the pond is carried along the bottom by the water to the drains where it is removed. This makes the pond virtually self-cleaning.

HATCHERY LIMITATIONS

Regardless of pond design the basic hatchery requirement is a good water supply. The water must be clean, free of pollutants, clear, and moderate in temperature. The ideal temperature is between 10 and 14° C, although few hatcheries on the Pacific coast have a constant supply of water at this temperature. Most hatcheries operate with water that ranges from 1 to 20°C. If, as sometimes happens, the water heats until it approaches the upper lethal limit of about 25°C, the fish are subjected to rapidly increasing stress and to warm water disease.

Once the water has passed through the tanks it may be directed to a filtering system, clarifier, or settling pond to remove waste materials before being discharged back into the adjacent stream.

High turbidity (too much sediment) decreases feeding efficiency and may cause gill damage if prolonged. The water should be fully saturated with dissolved oxygen, and the balance between oxygen and nitrogen must be maintained. To accomplish this the water is often passed through an aeration tower.

During rearing the fish may come in contact with a number of freshwater diseases. Some of these diseases may be transmitted from the female to her progeny in the ovarian fluid. To prevent this, all eggs taken during spawning are disinfected in a dilute iodine solution before being placed in the incubation room. Unfortunately, many of the diseases are transmitted through the water supply from other fish carriers upstream of the hatchery intake.

Disease may take many forms. Fungus may invade the eggs during incubation and is frequently a secondary infection in juveniles and adults. Viral diseases are less common and have no known cure. Bacterial diseases include: bacterial gill disease, kidney disease, furunculosis, and high temperature disease. Protozoan and metazoan parasites are also numerous, but they can be controlled by chemicals or by interruption of their life cycles.

As the fish grow they attract many predators. Birds and small animals such as mergansers, kingfishers, herons, gulls, mink, and otter find them easy prey. The larger the ponds, the more predators they attract. Nets over ponds, fences, noisemakers, trapping, and removal are all used to prevent fish being taken.

It is this protection from the uncertainties of the natural environment—the diseases, parasites and predators—and a supply of food far greater than

Kingfisher

could be produced naturally in a small stream that permit the production of large numbers of fish in a hatchery.

Because of the acceleration of growth rates and extended rearing programs, hatchery fish have been released in all months of the year. Ideally, smolts are released at a size equal to, or larger than, wild smolts, and at a time close to their normal peak migration. When they are ready, the screens on the pond outlet are removed and the fish allowed to migrate downstream. Within a few days they are in the estuary, where they remain for a period of time before starting their ocean migration.

Mink

FISH MARKING

The measure of the success of any salmon production scheme is the ocean survival of the fry or smolts produced and the contribution of these fish to commercial and sport fisheries.

Over the years many techniques have been used to mark juvenile fish so that they can be recognized when they reach adult size. The first widely used method was the removal of a single fin, usually the adipose fin or the right or left ventral fin. This was useful as long as estimates of survival were based simply on the relative returns of different groups of fish. But the fish do not have enough expendable fins to permit major studies on survival rates and ocean migrations, and the problem was further complicated by the fact that fish can sometimes regenerate a missing fin.

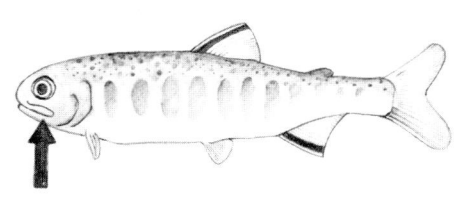

For more accurate results it was found necessary to clip at least two fins, and at one time or another all possible combinations of fin removals have been used. Even the outer extension of the right and left maxillary bones of the jaw have been removed in an attempt to find a suitable "mark." With the sole exception of the clipping of the adipose fin, all other fin removals or multiple marks have resulted in a significant decrease in survival.

Jaw clipping

A means had to be found to mark large numbers of juvenile hatchery fish that would not affect their survival. Marking techniques that had been developed for birds and small mammals were tried. Numbered metal clips have been applied to jaws, opercles, fins, etc.; metal or plastic tags on stainless steel wires have also been used. Whereas some of these devices are suitable for fish that do not grow much, they proved impractical for those whose size increased greatly.

Opercular tag

Branding has been used to mark fish. Early trials using hot branding techniques gave way to cold branding where a metal number or symbol was supercooled in liquid air and then applied to the side of the fish. The pigment cells touched by the brand were disrupted, and the outline of the brand could be seen. Unfortunately, the damaged cells often regenerated over a few months to make the scar indistinguishable when the fish became an adult.

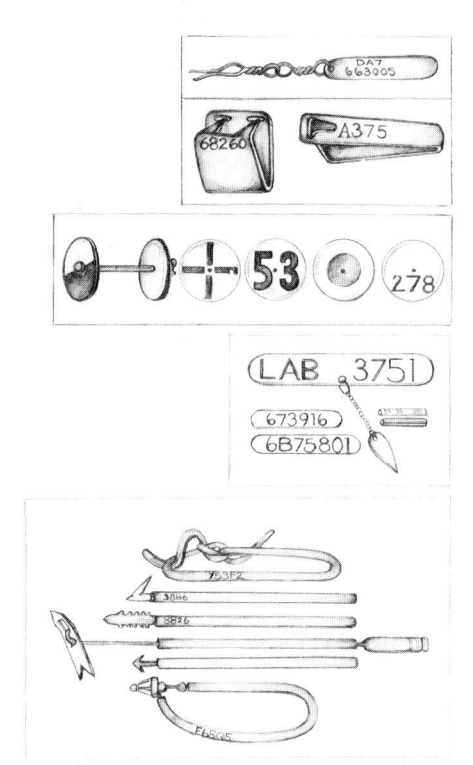

Opercular tags; Peterson disc tags; Internal tags; Spaghetti tags

Branding

Fluorescent dye marking

Cold branding as a short-term mark is still being used in the study of juvenile migration in large rivers, and for this purpose it is a useful technique.

The development of lasers revived interest in the branding of fish, but it does not appear now that a laser brand is significantly more permanent than the more conventional cold brand.

All branding techniques suffer the same deficiency: any external mark that is highly visible to a fisherman or to a fish sampler is also likely to be visible to a predator. Any fish that appears different from normal will be subject to a higher rate of predation. For this reason the results of hatchery experiments using branded fish are not the same as if the fish were unmarked.

To avoid predators being attracted to visible marks, juvenile fish were sprayed with a granular dye that becomes embedded in their scales and skin. Under normal light conditions the dye appears as a neutral color; however, when illuminated by ultraviolet (UV) light the dye fluoresces and becomes highly visible. This technique overcomes virtually all the objections to external marks but still presents two difficulties. Firstly, the distribution of the dye pigment on the body of the juvenile becomes scattered and to some extent lost when it reaches adult size. Secondly, recovery of marked fish by fishermen and others is more difficult when they have to view the fish under UV light. As a result this technique has seen only limited use.

Oxytetracycline is an inexpensive drug commonly used in salmon culture for the treatment of a variety of fish illnesses. It has been observed that when fish are given this drug in their feed, the areas that are in the process of being calcified—vertebrae, opercular bones, etc.—form a layer of material which, when exposed to UV light, also fluoresces. When fish have been given food with the drug and then food without the drug for alternating periods, a series of concentric rings about the centrum of a vertebra can be seen under UV light.

Although this is a very suitable internal mark, the method has not been used widely because of the difficulty in administering uniform amounts of the drug to all the fish in the sample, as well as the need for specialized equipment to detect the presence of the mark.

The most successful technique yet developed for marking chinook and coho smolts has been in use since the 1960s. A small piece of magnetized stainless steel wire bearing a visible external code is injected into the nose cartilage. To indicate the presence of this imbedded wire tag the adipose fin is removed. Amputation of the adipose fin does not result in any increase in mortality and has no apparent effect on the fish.

As the fish grows, this internal tag remains imbedded in the nose. When an adult chinook or coho salmon is caught that is missing the adipose fin, its head can be checked with a sensitive magnetic detector to determine if the fish has been tagged. If it has, the nose is amputated and the tag removed for decoding.

Initially, the tags were color coded with a series of up to six bands of color running the length of the tag and bound to the stainless steel wire in an epoxy coating. These have now been supplanted largely by binary-coded wire, where the code appears as a series of nicks in four lines at 90-degree intervals along the sides of the wire tag. The advantage is that more information can be etched into the surface of the wire. Each fisheries agency on the Pacific coast has been assigned one or more codes.

By international agreement only chinook and coho salmon bearing a coded wire tag are to be released with the adipose fin missing. Since 1973 a program has been underway with the co-operation of sport and commercial fisheries to recover large numbers of chinook and coho with missing adipose fins.

In 1974 more than 25,000 tags of Canadian and U.S. origin were recovered in British Columbia. More information about ocean distribution, and the numbers of fish from hatcheries in specific fisheries, was gathered in 12 months than had been gathered in the previous 12 years. By the end of 1975 almost 15,000,000 coded wire-tagged chinook and coho were released in the Pacific Northwest.

A small percentage of marked fish subsequently lose their tags, and a few fish either lose or are born without an adipose fin. For this reason not all adult chinook and coho salmon with missing adipose fins have a coded wire tag. The resulting minor errors are far outweighed by the general usefulness of the technique.

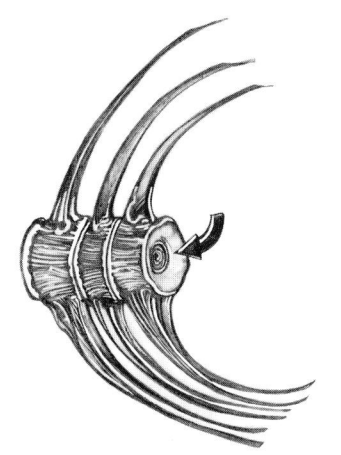

Oxytetracycline ring in vertebrae

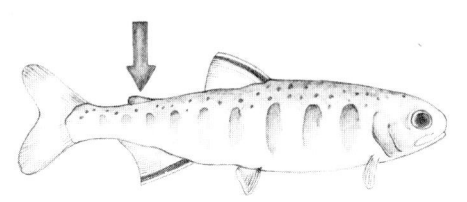

Adipose fin

Adipose fin clipping

Magnetic detector for nose tag

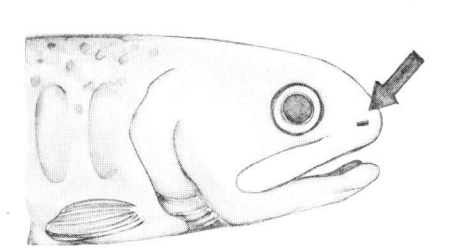

Binary-coded nose tag

Color-coded and binary-coded nose tags (1 mm)

MANAGEMENT

At one time, the salmon returning from the sea seemed an inexhaustible bounty. Now, modern salmon fishing fleets are capable of catching entire stocks. For this reason the total fishing effort is closely controlled. It is a complex task, involving many people who are, in greater and lesser degrees, "managers" of the fragile stocks. The role requires knowledge of the various stocks, their spawning times and places, their life cycles, migrations and environmental needs. It requires patient monitoring of the runs and judicious application of regulations to achieve the crucial balance between catches and escapement.

The manager begins by planning the fishing season. Prediction is the key. He uses the number spawned of the preceding generation as a gauge in forecasting the coming year's production. In this way the manager estimates the number of salmon expected to return as adults and sets a pattern of fishing that will permit sufficient salmon to reach the spawning grounds to maintain, and even increase, the stocks.

Things seldom work out as planned. On each day of the commercial fishery, reports are received on the number of fish caught, their age and size, and the numbers and types of fishing boats at work. Test fishing provides an index to the number of fish in the various areas. Fences are placed near the mouths of smaller rivers and streams, where the salmon are counted electronically to give an accurate record of migrating fish. Also, salmon can be caught and tagged at sea before they enter the fishery. Tags returned by fishermen help estimate catch rates.

The manager may apply one or more of these "tools" to assist him in determining if the stock size is as predicted. Or he may evaluate the fishing fleet—numbers, gear type, distribution, weather, and ocean conditions—for indications that total stock strength is not being confirmed by the catch. Many factors are involved. Tides affect migration rates, routes, and timing. Weather affects fishing. During bad weather, salmon may swim deeper, sometimes so deep as to be below the fishermen's nets. Also, fishermen cannot operate their gear as well during rough weather.

In some locations, luminescent plankton, disturbed by the nets, produce enough light to outline the nets at night. When this happens the fish can avoid the nets. They can also avoid nets made visible by coatings of mud and debris from rivers in flood. The mud-fouled nets may even be seen by the fish at night when the moon is full. On the other hand, turbidity sometimes obscures the net, making it more effective.

Mesh size is geared to a certain size of fish. If the fish in a run are much smaller than average, the gill nets will not be as effective, for the fish will swim through them. Fish holding in tight, visible schools are more easily caught by seines.

Counting fence

REGULATION

Once fishing has begun, the manager must determine whether he has to vary the proposed harvesting pattern or to allow fishing to go as planned. He has a number of options. To lessen the numbers being caught, he can restrict the type of fishing gear that can be used, decrease the number of days per week that fishing is allowed, or change the total fishing area.

Salmon often congregate off a river mouth in dense schools before going upstream. Here they are especially vulnerable to overfishing. Normally, then, to enable sufficient numbers to escape, sanctuary areas are designated off the mouths of salmon rivers and in certain bays and inlets along their routes. Outside the sanctuaries areas are open to fishing.

The manager can alter a fishing area by changing either its inshore or offshore boundaries. If he moves the inshore boundary out, the fishing area is reduced, the sanctuary area is increased, and the number of migration days' travel distance over which the fish can be caught is decreased. If too many fish escape, he may shrink the size of the sanctuary.

To control the catch by gear restrictions, the manager can regulate the gill net mesh sizes or net lengths. He is well aware, however, that it is costly for a commercial fisherman to have to change his equipment because of a new limitation and so uses this control sparingly.

Because salmon stocks pass through coastal fishing areas at predictable times, it might seem possible to alter the fishing period to accommodate either a greater or lesser abundance of fish. The disadvantage of doing this is that the catch should be taken over the whole of the run to preserve the variability of the entire stock.

MIXED STOCKS

In most salmon fisheries the area in which the catch takes place embraces the inbound routes for a number of stocks and species. If one stream in a river system produces salmon at a rate of six progeny per spawning pair of adults, and another stream in the system produces at a rate of three progeny, and both stocks migrate through the fishing area together, a problem arises. An acceptable catch rate for the one stock could, over a number of generations, depress the other stock to levels approaching extinction.

For this reason the fishery manager must find ways of fishing different stocks as independently as possible, even when they become mixed. The methods used are based on the salmon's migratory behavior and on selectivity by the fishing fleet.

If the inbound route and timing of each adult population can be ascertained, it may be found that some stocks use different migration routes, or are in a particular area at a different time from other stocks being fished. With this knowledge the time and location to harvest each stock selectively can be determined. Such information is usually obtained from tag recoveries in commercial, recreational, and native food fisheries, as well as on the spawning grounds.

Another way is to set up a selective fishery so that some stocks can be fished more heavily than others. Selection is usually by size, but fishing by species and sex is also possible. With gill nets, large species such as chinook can be harvested while sparing smaller species such as sockeye. Similarly, individual stocks of the same species which vary in size may be harvested selectively. As there is often quite a difference in size and shape between the sexes, gill nets can be used to select for sex.

In management of a size or sex selective fishery, ending up with an imbalance among the spawning stocks can present a problem. Also, if there are changes in size composition during the fishery—for example, large fish arriving early and small fish late—the rate of escapement throughout the season may not be uniform.

By using one or more methods for selectively exploiting stocks, the fishery manager is often able to overcome the problems of mixed stock fishing.

WATERSHED CONSERVATION

Most people coming to the Northwest in the early days sought a living in logging, fishing, or mining. Because of the vast grassland prairies, the Rockies, and other mountain range barricades, pioneers usually came by way of the sea. The rivers were the access routes, first for the canoe brigades of the fur traders, then for the packhorse trains of the miners. These gave way to the road and rail networks that—like the brigade trails—followed the river valleys. The loggers used the rivers for log drives; the miners used them for hydraulic and placer mining.

During these years the resources of each watershed were treated separately by the different industries, and this led to conflicts in resource management, since salmon, timber, and mine sites are all integral parts of a watershed. The problems were familiar to forest and fisheries managers, but it was not until the 1970s that their concerns were shared by the general public, especially about the future of the salmon and steelhead trout. Conservationist fears that logging activities were inflicting permanent damage on salmon and steelhead fisheries led to an inquiry in 1973 by the British Columbia legislative standing committee on forestry and fisheries. The inquiry dealt with the problems of protecting the stream banks and lake shores, as well as the whole question of resource management.

The first major attempt in British Columbia to resolve the conflict of the salmon versus logging and mining interests was made in the Skeena River watershed—an area spanning 10 000 sq km (4,000 sq mi)—by mining and forestry companies working with government agencies toward integrated resources management. Thus far, this program has been successful. Open-pit copper mines have elaborate waste control systems to prevent heavy metal ions from escaping into salmon streams and lakes. Forestry firms bundle their logs with steel strapping for towing across Babine Lake. Both mining and forestry companies employ "environmental managers."

The Skeena watershed program is an example of the "multiple resource use planning" that has evolved in recent years between the two major industries in British Columbia (forestry and mining) and the government agencies charged with the protection of the land and its creatures. The B.C. Forest Service is responsible for the woodlands, and the salmon come under the Canadian Department of Fisheries. Since the mid-1960s the Forest Service has routinely advised Canadian Fisheries of impending long-term logging contracts. More recently it introduced the "resource folio system" in which all resource values of an area are considered in the long-range planning of timber operations.

In 1971 the British Columbia legislature passed the Environment and Land Use Act. Now all new major development must be preceded by studies of its effects on other resources. In addition to this, all projects involving federal funding or federal lands come under the Environmental Assessment and Review Process. For these reasons little is undertaken in British Columbia today—highway; oil exploration; mining; dam construction—that has not been reviewed by the Canadian Fisheries Department for its possible effect on the salmonids or their habitat.

LAKE ENRICHMENT

Experiments in the fertilization of normally unproductive sockeye nursery lakes have been underway on Vancouver Island since 1970. The idea is not new. For centuries European and Asiatic fish farmers have fertilized the ponds in which they raise fish. But the sockeye salmon of British Columbia are not the easy-to-rear carp, catfish, and yellowtail species raised by Old World fish farmers. A 50-km (30-mi) long, 200-metre (630-ft) deep mountain terrain lake is far different from a farm pond.

Young sockeye spend one or more years in a freshwater lake before leaving for the ocean. Their chances of reaching maturity in the sea are related to their size as smolts when they leave the nursery lake: the larger they are as smolts, the greater the numbers that will survive to return as adults.

Smolt size and abundance depend on the availability of food in the nursery lake. The young sockeye eat zooplankton—particularly tiny crustaceans—that feed in turn on phytoplankton: microscopic plant life. These minute plants form the base of the aquatic food chain. To grow and multiply they require light, as well as dissolved minerals and other basic nutrients. If nutrients are scarce, phytoplankton will be reduced and every link in the food chain diminished.

Nutrients arrive in a lake with runoff waters from the lake's drainage basin. But in the high rainfall areas of British Columbia, lush plant growth depletes soil nutrients, so lakes on Vancouver Island and along the mainland coast are often deficient in dissolved nutrients. Where logging has stripped the terrain of trees, runoff water does not remain on the land long enough to leach out minerals. Without trees to slow it down the rushing water carries large amounts of sediment into the lakes. This material sinks quickly to the dark reaches of the bottom, where the nutrients are largely unavailable to the phytoplankton.

The depth of a lake directly affects its production of food organisms. In shallow lakes, wind action and the altered density of the surface water due to seasonal temperature fluctuations result in mixing of the water mass from top to bottom. In deep lakes, only the upper layers of water are mixed while the depths remain constantly cold and still. Nutrients tend to become locked into the bottom materials or to be stranded in solution in water that mixes little, if at all, with the warmer productive water layers nearer the surface.

In a deep lake the limited supply of nutrients found in the upper layers is greatest in the spring and early summer, the time when increasing light and higher temperatures stimulate the winter-dormant phytoplankton back into production. There develops a succession of microscopic plant communities (each capable of existing on lower concentrations of minerals and nutrients than its predecessors) along with parallel associations of animal life (zooplankton) that feed on the plants. Some recycling of decaying plant and animal matter occurs, but much of it drops to the bottom, and the nutrients it contains are no longer available to the organisms in the surface waters.

In sockeye nursery lakes the carcasses of spawned-out salmon are also a source of primary nutrients, a source that is reduced drastically by the commercial fishery, which annually eliminates most of the returning adults before they reach fresh water. Fisheries biologists speculated that the sequence of heavy salmon harvests, fewer spawners, fewer nutrients, fewer food organisms, and thus fewer sockeye smolts, could be one reason for the general decline of sockeye along the Pacific coast. That line of conjecture led to the suggestion that the missing nutrients might be replaced artificially. Great Central Lake on Vancouver Island was chosen for the experiment.

Great Central Lake lacked nutrients and had been relatively unproductive of salmon. Year-old sockeye smolts leaving this lake for the sea were much smaller than sockeye emigrating from Babine Lake on the mainland, which had a nutrient content two to three times as great. Steep shorelines and narrow shallows also recommended the lake for the experiment: there was little area occupied by rooted aquatic plants that might take up fertilizer intended to nurture the free floating plankton. In addition, the limited variety and lack of plankton species would make it easier to detect any changes that might follow fertilization. Finally, the minor commercial fishery for Great Central Lake sockeye stocks could be easily monitored. Beginning in 1970, from June through October, five tons of fertilizer a week—100 tons in all—were released into the lake. The fertilizer consisted of a commercial grade of ammonium phosphate and ammonium nitrate, plus trace elements and a small quantity of organic solubles. The dissolved nutrients were released in the wake of a boat.

The fertilizing operation was timed to begin when the lake had stratified, that is, when the upper layer of lake water had warmed and become less dense, forming in effect a separate lake with its own circulation, which floated on the cold dense water of the depths. The upper layer, the "epilimnion," is the biologically productive area in a deep lake. The lower layer, or "hypolimnion," lacking light and sometimes deficient in oxygen as well, may be nearly lifeless. Development of the epilimnion coincides with the highest concentration of nutrients and the lengthening hours of daylight that stimulate the spring bloom of phytoplankton. These optimal conditions for plant growth can result in microscopic plants doubling their numbers every few hours. Zooplankton, the tiny animals feeding on the plant life, thrive. Normally, the period of rapid growth subsides in six to eight weeks, and the amount of phytoplankton available to the zooplankton diminishes as the available nutrients are consumed. By adding fertilizer to the lake waters researchers hoped to maintain the early spring level of nutrients over five or six months, thereby sustaining the greater plant and animal life for the same extended period. With more zooplankton to feed on, more sockeye fry could be expected to survive and to attain larger size as smolts.

The amount of fertilizer added to the lake was calculated to keep the nutrient level at or slightly below the natural level for early spring. Too much fertilizer (particularly phosphate) might have produced the symptoms of an overenriched, or eutrophic, body of water, perhaps including a rapid increase in undesirable algae in the depletion of oxygen. The application of the fertilizer in small batches at weekly intervals (rather than in a single 100-ton drop) was designed to avoid overenrichment; it was believed, too, that one very large treatment would quickly dissipate and fail to boost the food chain for sufficient time. Applying the fertilizer over a period of weeks also gave the researchers better control over the experiment and provided opportunity to make changes or adjustments as the effect of the earlier fertilizing became apparent.

The results of the 1970 experiment were encouraging. A phenomenal eight-fold increase in the abundance of zooplankton occurred in Great Central Lake, and the growth rate of sockeye fry was doubled. The program was repeated for four consecutive years, with comparable results. Soluble nutrients added in small, frequent applications were found to be entirely taken up by the plankton within a few hours of application, the greater part of the fertilizer being absorbed within minutes. No persistent algae blooms occurred, and, although there were temporary changes in water clarity owing to the increase in microorganisms and also some changes in the diversity and relative numbers of the plankton species present, the changes did not persist when fertiliztion was stopped. Apparently, artificial enrichment does not impose irreversible change, a vital consideration when experimental procedure is designed to modify a natural environment.

In the years since the enrichment program began at Great Central Lake, the average annual returns of adult salmon have increased to 20 times those of the average in pre-fertilization years (50,000). In 1977 over a million sockeye returned. With such positive results the experiment was continued and the program broadened to include other nutrient-deficient lakes. Four other Vancouver Island lakes—Henderson, Hobiton, Mohun and Kennedy—as well as Long Lake on the mainland are now subject to enrichment.

When the experiment began, a boat was used to spread the fertilizer, but this was expensive in manpower and equipment. It was realized that access roads were needed, and crews would have to be stationed at each lake all summer. The logistical problems of trucking bags of fertilizer to these lakes, of providing storage for it, of mixing it, and, finally, the frequent boat trips required to carry out the weekly treatment, prompted the planners to try aerial application. Once a week now throughout the summer, a four-engined DC6B transport plane modified for spray application flies low over each lake, trailing a plume of liquid fertilizer.

Mohun Lake, north of Campbell River on Vancouver Island, is still being fertilized by boat. Federal Fisheries and the B.C. Fish and Wildlife Branch are co-operating to increase the numbers of kokanee—landlocked sockeye—in Mohun Lake. The fertilizer is mixed with water in Campbell River and trucked to Mohun Lake, where it is distributed in the wake of a boat.

Because of the differences between lakes, sockeye returns following lake fertilization may vary widely and unexpected results can occur, as they can with any manipulation of complex natural systems. To ensure that these results are positive, and to avoid unpleasant surprises, careful controls are included in the lake enrichment procedure.

STREAM ENGINEERING

It has long been recognized that the freshwater streams of the Pacific coast drainage form the base of the salmon resource. Thus far, much of industrial man's impact has been negative, but it remains true that almost anything done to improve the streams will benefit the salmon.

As a means of regulating waterflow, a check dam installed at the outlet of a headwater lake can be most effective. It prevents destructive floods in times of heavy rains and guarantees flows in times of drought. Alternatively, supplemental waterflows can be arranged by redirecting a shallow stream into a narrow section of the channel. Sometimes needed water can be diverted from one watershed to another.

To help the salmon, spawning streams can be cleaned and deadfalls and beaver dams that impede adult spawning runs can be cleared out. Ironically, such obstructions are not entirely detrimental; they slow stream runoffs, create pools needed by the young of coho and steelhead, and provide cover for both salmon and trout.

Log jams, landslides, and waterfalls cause death or injury to leaping fish and delay the timing of runs. Impassable blocks force the salmon to spawn downstream, which results in crowding of the accessible lower spawning grounds while the upstream gravel beds go unused. Such blockages are breached or bypassed, using dynamite or a bulldozer if necessary.

The health of the stream and the life within it depend on the streamside vegetation. Shade is essential. Without shade, sunlight can cause lethal water temperatures for salmonids. Another effect of direct sunlight on a stream is the excessive growth of filamentous algae. Plant life can become so thick that fish are unable to swim unimpeded. The death and decay of this plant life, in the process of eutrophication, rob fish of oxygen. Thus the planting of trees in exposed areas and along banks does much to improve streams. Not only does the shade inhibit the growth of algae but also it provides welcome cover to both spawners and offspring fry. Overhanging willows, for instance, shelter salmon and trout feeding in quiet pools.

Log jam

Streamside flora reduces sedimentation. The fibrous networks of roots stabilize the banks by slowing erosion caused by the undercutting water. Unchecked, erosion can choke the gravel spawning beds with smothering silt. But trees and shrubs do more than just stabilize the banks and provide shade and cover. Leaves are the major source of food materials in forested streams. The litter which falls into the water in autumn decomposes during the winter under the action of fungi and bacteria. This microflora is the food of many aquatic insects which in turn feed coho, chinook, and steelhead juveniles.

Deciduous litter has more food value than coniferous. It also sustains more insects. In late summer, when streams are at their lowest and aquatic foods are scarce, young salmonids may subsist on flying insects. The numbers of such insects are directly influenced by the climate. Many more aphids and midges alight on sunlit water than in heavily shaded areas.

Stream "management" implies man-made changes. Planking, logs, or concrete can be used to divert the waterflow from eroding banks, thus reducing harmful sedimentation. These same materials can also be used to improve the pool-riffle sequence of salmon nursery streams. A combination of riffles and pools of roughly equal areas is ideal. The pools should be deep but not long; small pools support more juvenile fish than do large ones.

In normal streams, wing deflectors can be built from the sides. They narrow the channel, deepen the pools, and emphasize the sinuous course of the river.

In steep gradient streams, low barriers can be installed across the flow. These weirs accentuate the "steps" and deepen the pools below the artificial ledges. Steelhead trout find shelter below such barriers and use the standing wave of turbulence to assist their leap to the next higher pool.

Instream devices have limited application, often being impractical in the high-velocity freshets that drain much of the Pacific west coast and in rivers with widely fluctuating rates of flow and rapidly shifting gravel bars. Their usefulness is in small tributary creeks or wherever some measure of flow control is available.

FISHWAYS

Waterfalls and rapids are the most common obstructions encountered in the coastal streams of British Columbia. Fisheries personnel classify such natural blocks as either "impassable" or "partial." Impassable obstacles deny all salmon access to the upper river; partial ones cause injuries (or death) and delays which can impair spawning efficiencies.

Sometimes it is possible to provide upstream passage with the judicious use of explosives and rock removal equipment. Often, a fishway must be built. This concrete and steel structure dissipates the violent energy in the water while allowing fish to surmount the obstruction without stress.

There are three major types of fishways used in British Columbia. The weir type consists of a series of vertical partitions installed laterally at intervals down the length of a specially constructed channel or flume. The water drawn from the river upstream of the obstruction flows over the tops of the successive partitions—each slightly lower than its predecessor—creating a series of step-like pools which salmon can ascend with ease. Weir-type fishways are most effectively used where water levels remain fairly constant. Where water levels fluctuate, the weir fishway requires constant attention. Each time the streamflow—and thus the water depth—changes, adjustment of baffles or stop logs is necessary.

Leaping salmon

The vertical slot fishway also creates a series of pools and drops, but the water flows from pool to pool via one or more narrow slots which extend to the full height of each partition. The vertical slot type is self-adjusting; that is, it provides the fish with passage through the ladder at all levels of water in the river. As most coastal streams are subject to sudden spates, and given the relative isolation of most fishways, this is a particular advantage.

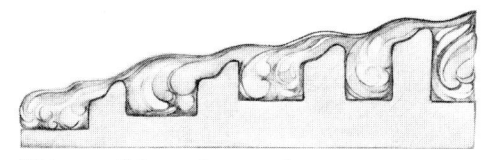

The denil type fishway is, essentially, a short section of flume in which baffles of various shapes and sizes have been affixed to the sidewalls and floor. The energy of the water passing through the central open section of the structure is dissipated in turbulence caused by the baffles, leaving a narrow corridor of low velocity flow through which salmon can ascend without difficulty.

Weir-type fishway (cutaway)

This type is ideal for helping smaller runs of salmon circumvent blocks where more expensive fishways are not warranted. Also, denil fishways can be installed at much steeper gradients. For a given height of obstruction, they can be shorter. This economy is reflected in cheaper construction costs.

A fishway's entrance has strategic importance. Fish are attracted to the area of greatest flow, which is usually over the falls. If the fishway entrance can be positioned close to the falls, or in an area where the fish will congregate, the number of their unsuccessful attempts to mount the falls will be reduced. Sometimes when an attractive entrance location cannot be found, rock work, a concrete wall, or even a barrier dam is used to divert water and fish toward the entrance.

Fishways are not new. They have been in use for about 300 years in Europe and were employed in British Columbia in the early 1920s and '30s. These structures were all of the weir type and did little to help fish over waterfalls, rapids, or other natural impediments. It was not until the evolution of the vertical slot baffle, and its first employment at Hell's Gate in the Fraser Canyon, that the fishway was firmly established as a tool in the management of salmon stocks.

As salmon runs to the areas above Hell's Gate began to increase, studies of the fishways revealed that when the water levels became higher or lower than the range they were designed for, delays to migrating stocks occurred. Three more fishways were built on the east bank and one on the west bank between 1947 and 1966 to facilitate passage for the fish. Fishways now operate over the full range of river levels (3 to 28 m; 10 to 92 ft) that occur in the Fraser during the salmon migration period. These fishways at Hell's Gate, coupled with a series of smaller fishways at the Bridge River and Yale rapids, have allowed a substantial restoration of both pink and sockeye runs in the Fraser.

The success of the Hell's Gate fishways triggered a program of construction by the Canadian Department of Fisheries. Starting in 1951 with two fishways at Moricetown Falls on the Bulkley River, installations have continued at a rate of about one per year. Almost all these fishways were built at partial obstructions where passage was made difficult because of water levels at migration time. To evaluate the success of some of these structures, tagging studies were conducted before and after construction.

At Meziadin, in the Nass River system, for example, counts show that 70,000 sockeye died at the falls in 1966, the year prior to the building of the fishway. Two years afterward, the runs were ascending through the fishway with no difficulty. In those two years the total stock rose from an average of 240,000 to 403,000 sockeye. The increase more than offset the total cost of the fishway ($750,000) in the first two years of returns.

Concrete fishways built in rock excavations are expensive. This, plus the high cost of providing construction access to the site, often renders such projects uneconomic. Prefabricated denil fishways suitable for helicopter air lift have been installed at remote rapids and waterfalls where access costs were hitherto prohibitive.

One such steep pass fishway was manufactured of aluminum in Vancouver and transported by vessel and helicopter to a remote site on the Kakweiken River, near Knight Inlet, where it was installed at a waterfall. Before this time pink salmon had never been seen above these falls. Today, pink salmon spawning runs of 100,000 to 230,000 annually ascend the falls to use a previously unused streambed. Now that it is known that these salmon will deploy above the falls in large numbers, construction of a permanent concrete fishway becomes economic. Thus the portable fishway, useful where anticipated returns do not justify the cost of an expensive permanent structure, can also be used as a preliminary testing device.

Fishways are found to be less useful in association with power dams, as illustrated by the Columbia River experience in Washington and Oregon. Dams along the main stem of the Columbia incorporate fishways in their structures. Each fishway constitutes an obstacle to the salmon that results in delays to their migrations. The salmon must first find the fishway, ascend it, then grope its way through the still water impounded behind the dam until it finds the stream it seeks.

Temperatures in reservoirs and low-water parts of the river system are often higher then the 15°C upper limit for healthy salmon. Disease such as that produced by the bacterium *Chondrococcus columnaris,* or "high temperature disease," is recognized as one of the major factors associated with the decline of salmon runs in the Columbia River system. In addition, the main river and some of its tributaries at times become supersaturated with dissolved oxygen and nitrogen produced by the plunging of water over dam spillways. Supersaturation at levels of 110 percent and more can be lethal. Many dead adult chinooks found along the Columbia have shown symptoms of the bends.

In recent decades, measures have been taken to stabilize the salmon populations on the Columbia River. To compensate for spawning grounds lost by dam construction and the blocking of migration routes, fishways have been constructed and obstacles removed. At some dams, hatcheries have been installed and whole stocks of salmon propagated artificially.

For mostly economic reasons the raising of salmonids from egg to pan-frying size has not been highly successful in North America; it has been cheaper to catch wild fish in the traditional ways. But as the numbers of wild salmon decline we are relying more on artificial propagation, principally in hatcheries and spawning channels.

AQUACULTURE

Thanks to the research done in these large plants and the experience gained from it, salmonid farming is becoming viable. The protected bays, inlets, and straits of the Pacific coast, along with the mild climate and abundant rainfall, provide an ideal environment for such enterprise. There is protection from storms; there are stable water temperatures and strong tidal currents to bring in dissolved oxygen and nutrients. In addition, salmonids are easy to breed and rear, they grow fast, and they bring a good market price. Feed for penned salmon is commercially available and relatively inexpensive.

Since the fish farmer must start with a hatchery process to incubate the salmon eggs, an assured supply of clean fresh water is of paramount importance.

On leaving the hatchery the fry are reared in tanks, ponds, or raceways to seagoing smolt stage before being transferred to salt water where they are held to marketable size. The build up of waste food, fecal matter, and other detritus in holding tanks, raceways, or pens must be avoided or outbreaks of disease will occur. Prevention of disease is easier than trying to treat infected fish, so good fish farmers try to maintain an adequate flow of fresh clean water, avoid overcrowding, minimize handling, provide a balanced diet on frequent feeding schedules, and observe high standards of sanitation.

Although the floating fish pen had been used for some time in fresh water, the Japanese were the first to recognize its potential for marine fish farming. They use cages of synthetic netting supported on bamboo frames that are floated on oil drums. Lines from each of the bottom corners of the cages are attached to weights on the sea bottom; these, and the bamboo frames, maintain the shape of the cage. The fish cages are not operated singly but in groups of ten or more. Walkways on the framework between the rows give workers access to the pens. The fish thrive on a diet of chopped fish or shrimp wastes.

The floating fish pen has many advantages over the fixed enclosure. Free from tidal changes, small units can be used, each amenable to cleaning, fish sorting, and feeding. The fish farmer can change the position of the small cages to offset fecal pollution which can occur in shallow areas. Finally, flushing with deeper water below the pen is possible by the use of compressed air "bubbler" systems similar to those used to keep canals and waterways ice-free and open to boats in winter.

These are the advantages, but floating fish pens have some drawbacks. They are susceptible to extremes of weather, such as storms with damaging waves. Seaweed may clog the netting. The pens have to be regularly dismantled, and the netting cleaned, disinfected and hung up to dry. The positioning of the pens may cause the fish farmer legal problems, for public access, water rights, obstructions to navigation, environmental impact, and zoning bylaws can all be matters of legal dispute.

And there are the inevitable predators. On land, the tanks, ponds, and raceways have to be protected from both bird and animal predators. Offshore, a covering of netting protects the penned salmon from diving birds and otter, mink, and seals, but dead salmon that sink to the bottom of the pens incite voracious dogfish sharks. These razor-toothed carnivores tear holes in the nets, allowing salmon to escape. To prevent this from happening heavy gauge double-walled nets for the saltwater pens are sometimes used.

The hazards in aquaculture are similiar to those facing farmers everywhere: predation, disease, and weather threatens the stock; high overhead and low profit margins threaten the enterprise. Where permitted by law (California, Oregon, and Alaska), some private operators have begun "ocean ranching." They incubate salmonid eggs (primarily chinook, coho, and chum) and release the subsequent smolts. By greatly increasing the numbers of juveniles going to sea, the ranchers hope to harvest sufficient adult salmon returning to the home stream some years later to make the venture pay.

ENHANCEMENT

The fisheries manager has the means to increase, or "enhance," the salmon stocks through the use of such tools as hatcheries and spawning channels. As with most tools, improper application can be dangerous. Enhancement projects can create severely mixed stocks and so cause further management problems. The danger is that of decimating the natural stocks. Although hatchery-produced salmon could withstand 95 percent being taken, how is the fisherman to separate wild salmon from their artificially bred relatives? It is possible to use technology safely in a number of ways. Fishways and stream engineering can open up new spawning and rearing areas. Stocks that can be fished independently of other stocks can be enhanced; those threatened by overfishing or environmental problems can be rehabilitated.

The techniques must be used with caution lest they create more problems than they solve. Stocks considered for enhancement—especially transplants—must be free of disease to avoid the loss of both natural and enhanced stocks.

Likewise, stocks to be enhanced should not conflict with other species in the system. For example, multiplying the stocks of coho or steelhead may result in greater predation of pink, chum and herring juveniles, with resultant decreases in their populations.

The scale of enhancement must not exceed the natural limitations at any later stage of life. Although chinook smolt production could be expanded by many hundred times, it would be pointless to do so if the estuary in which the chinook juveniles must rear could not support that number.

Salmon enhancement will be futile if Canada has no control over the foreign fishing fleets which threaten offshore and inshore fisheries. It was to assure management control over the resource that Canada, in 1977, extended its jurisdiction of fisheries to 200 miles. The decision to do so was based on the conclusion that the only way to ensure effective conservation is for Canada itself to assume management responsibility.

Authorities have been aware for some years of the need for such a program, both to ensure the survival of the Pacific salmon and to increase their numbers. It is only now, however, that success has been assured. The productive capacity of the stocks is becoming known; management methods are more exact. Techniques to multiply the various species have been developed in production facilities.

Nevertheless, planners are wary. They know the mistakes of earlier generations of fish culturists. They realize the uncertainties inherent in complex ecological systems, especially when disturbed by man. Today's fishery planners regard each salmon project as an experiment. And they will not undertake a project if the scientific means to evaluate it afterward are not available.

They know that decisions taken today often foreclose options needed tomorrow. They are aware that large capital investments commit them to policies that try to recover such costs. The larger the investment, the more pressure there is to "make it work" by pumping in more and more capital.

If the facility was a poor idea in the first place, or if some climatic, biological, or market disaster intervenes to negate its usefulness, political reasons may prevent its being closed down.

Ocean Distribution of Six British Columbia Salmonid Species

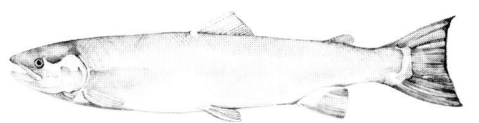

Steelhead trout reach 114 cm (45 in). Juvenile steelhead prey on young salmon

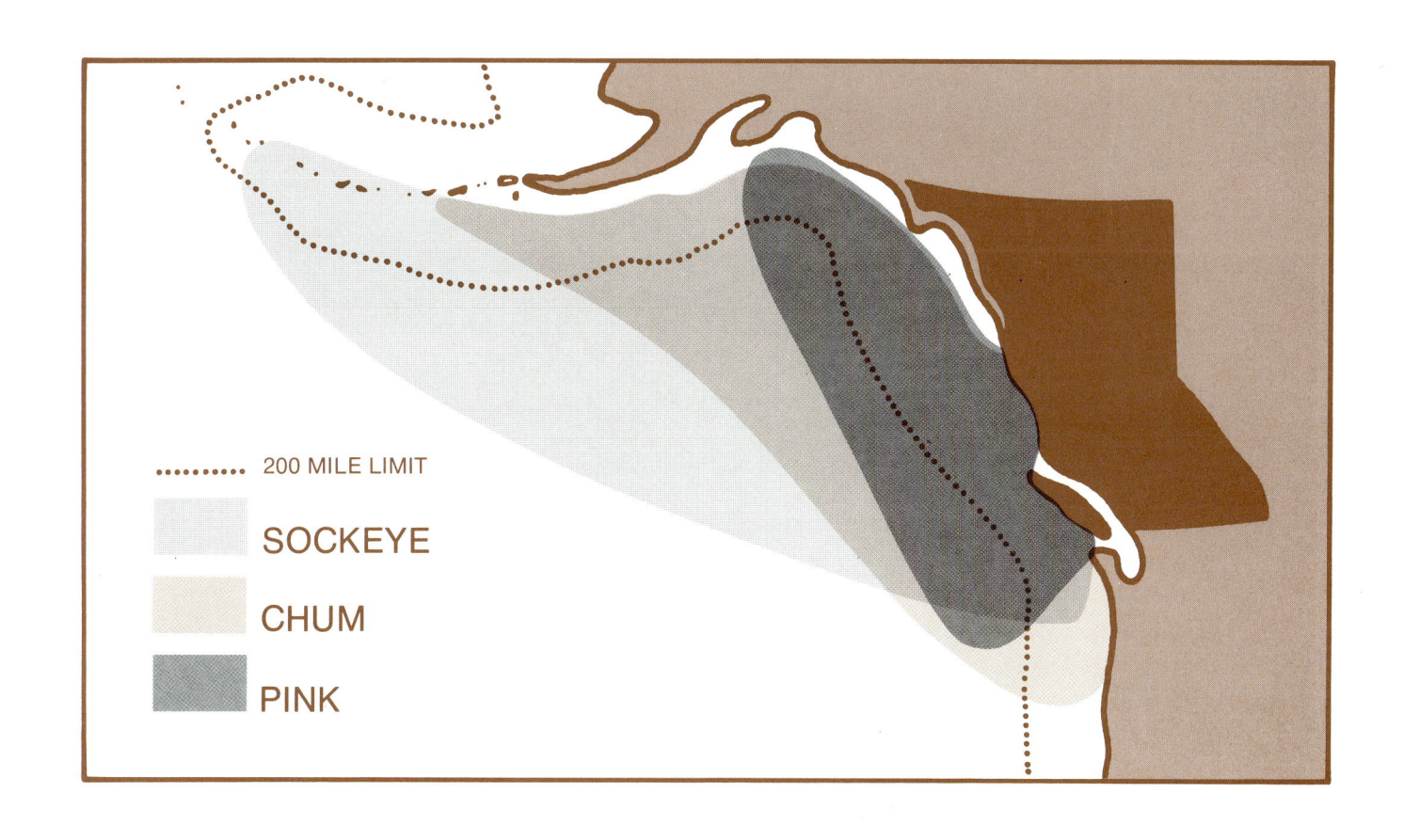

.... 200 MILE LIMIT

SOCKEYE

CHUM

PINK

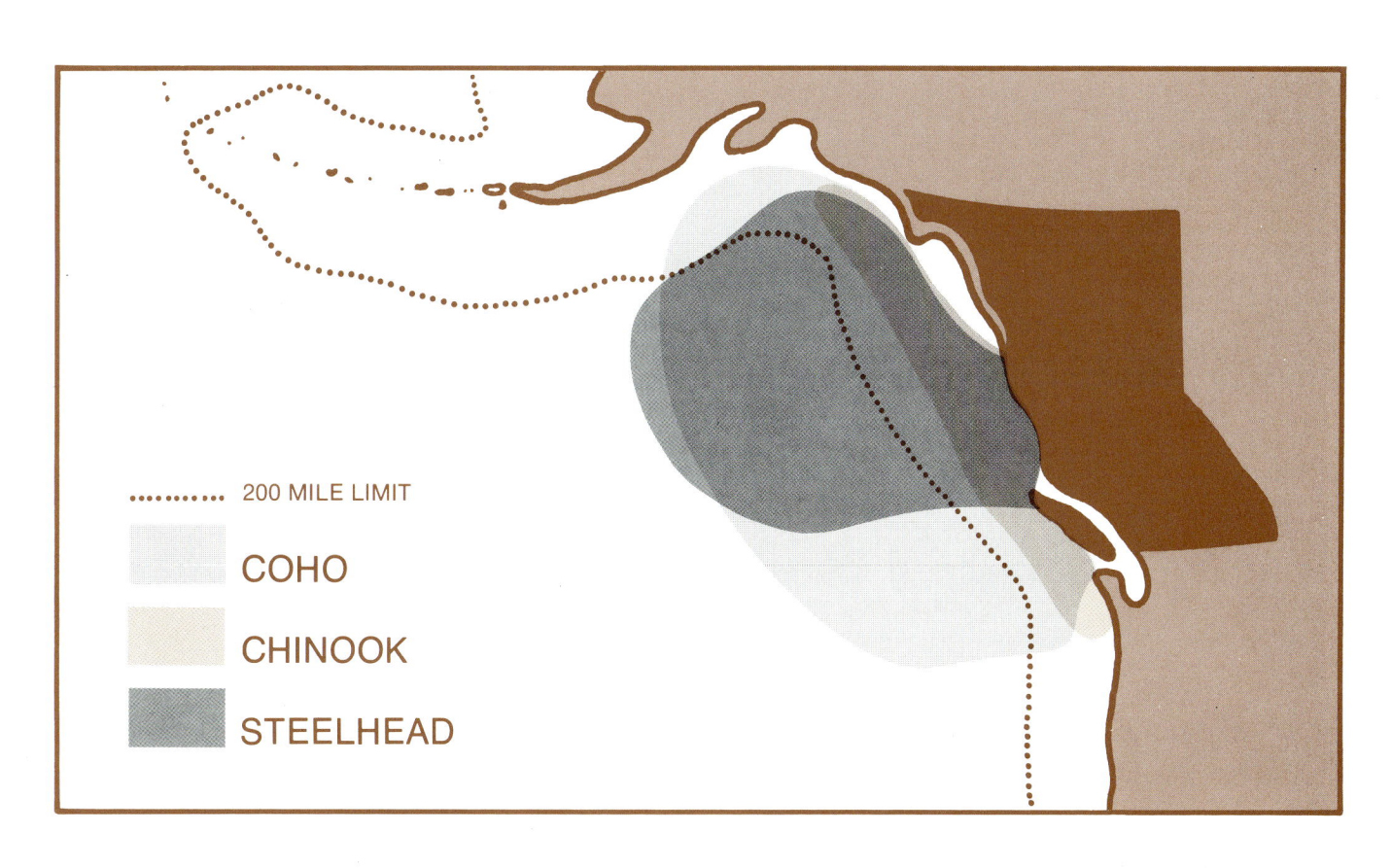

.... 200 MILE LIMIT

COHO

CHINOOK

STEELHEAD

Salmonid Program planners are aware that uncontrollable circumstances could—and probably will—arise. A new source of pollution in a major watershed could decimate salmon populations before it could be detected and controlled. Several successive drought or flood years could occur. The international treaty system could fail, resulting in overfishing by foreign fleets on the high seas. Disease organisms, a chlorine spill, or some other agent could wipe out a hatchery's production.

The history of salmon culture on the northwest coast is not one of conspicuous success. What is clear is that it is not easy to manipulate the salmon stocks exactly the way man wants. Our efforts have only succeeded in part. Uncertainties abound. For instance, vagaries of climate prevent us from accurately predicting salmon abundance from year to year. Because the effects of conditions in nature (and man's interferences) are multiplicative, there is a wide variation in returns for any number of spawning adult salmon. The implications of enhancement programs are that results may be discouraging for a series of years but then an explosive success may make up for years of poor returns. The process is akin to a gambling game, except that with enhancement the odds are shifted so that there is a smaller chance of losing and a bigger chance of winning.

Some Enhancement Projects in British Columbia

LEGEND

SPAWNING CHANNEL
1. Big Qualicum River
2. Little Qualicum River
3. Puntledge River
4. Jones Creek
5. Weaver Creek
6. Seton Creek
7. Pinkut Creek
8. Nadina River
9. Fulton River
10. Gates Creek

HATCHERY
1. Big Qualicum River
2. Robertson Creek
3. Puntledge River
4. Quinsam River
5. Capilano River
6. Fulton River
7. Abbotsford

JAPANESE STYLE HATCHERY
1. Bella Coola
2. Pallant
3. Thornton
4. Tlupana

FLOW CONTROL
1. Big Qualicum River
2. Wolf Lake
3. Pinkut Creek
4. Fulton River

INCUBATION BOX

FERTILIZATION
1. Hobiton Lake
2. Long Lake
3. Henderson Lake
4. Great Central Lake
5. Kennedy Lake

FISHWAY
1. Maggie
2. Skutz
3. Sproat
4. Stamp
5. Karmutsen
6. Quatse
7. Yale Rapids
8. Hell's Gate
9. Sakinaw Creek
10. Bridge River
11. Farwell Canyon
12. Kakweiken
13. Kadjusdis
14. Moricetown
15. Naden
16. Meziadin
17. Indian River
18. Embley
19. Fairfax

REARING POND
1. Big Qualicum River
2. Puntledge River

1. Inches Creek
2. Blaney Creek
3. Atnarko River

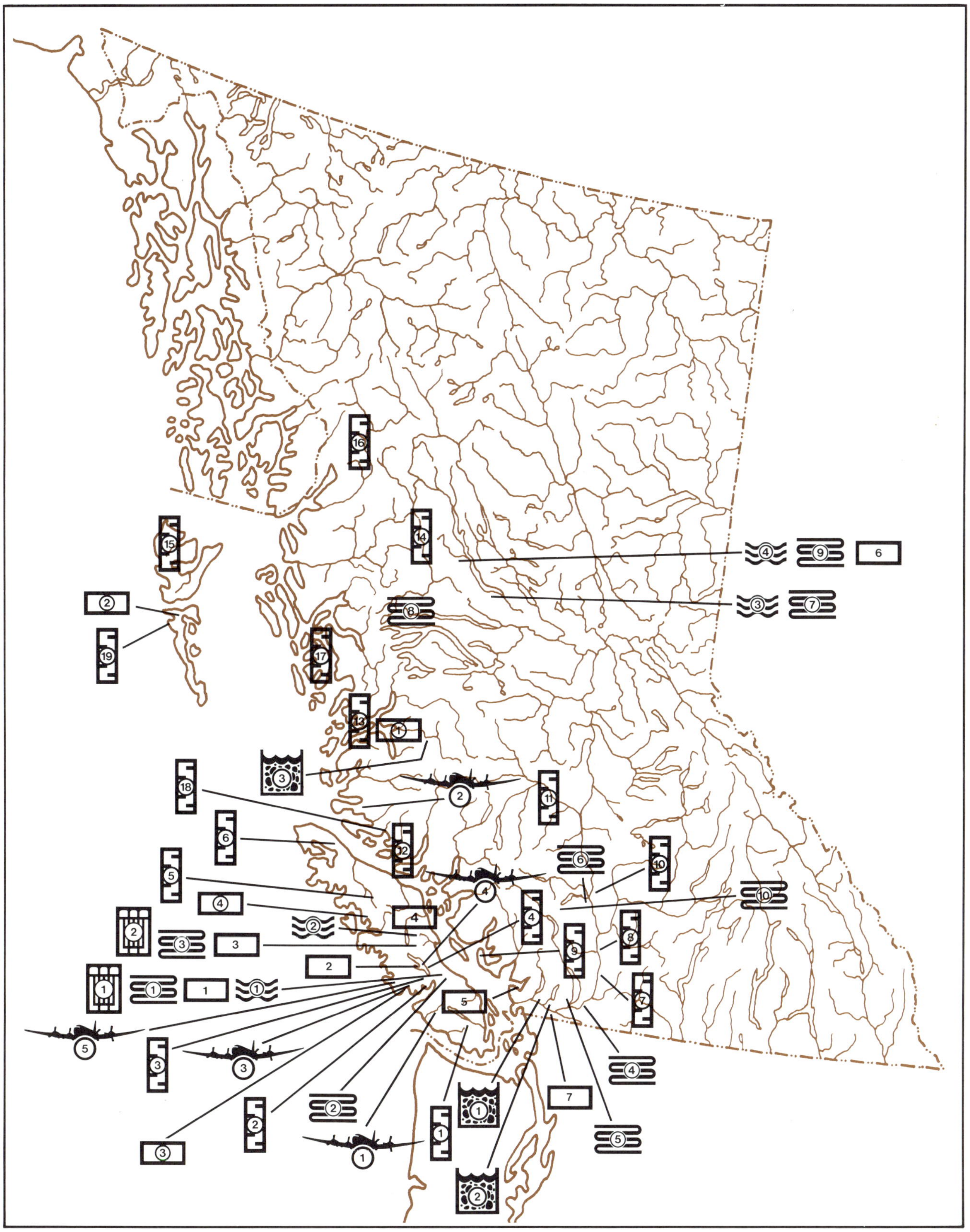

The Salmon Run

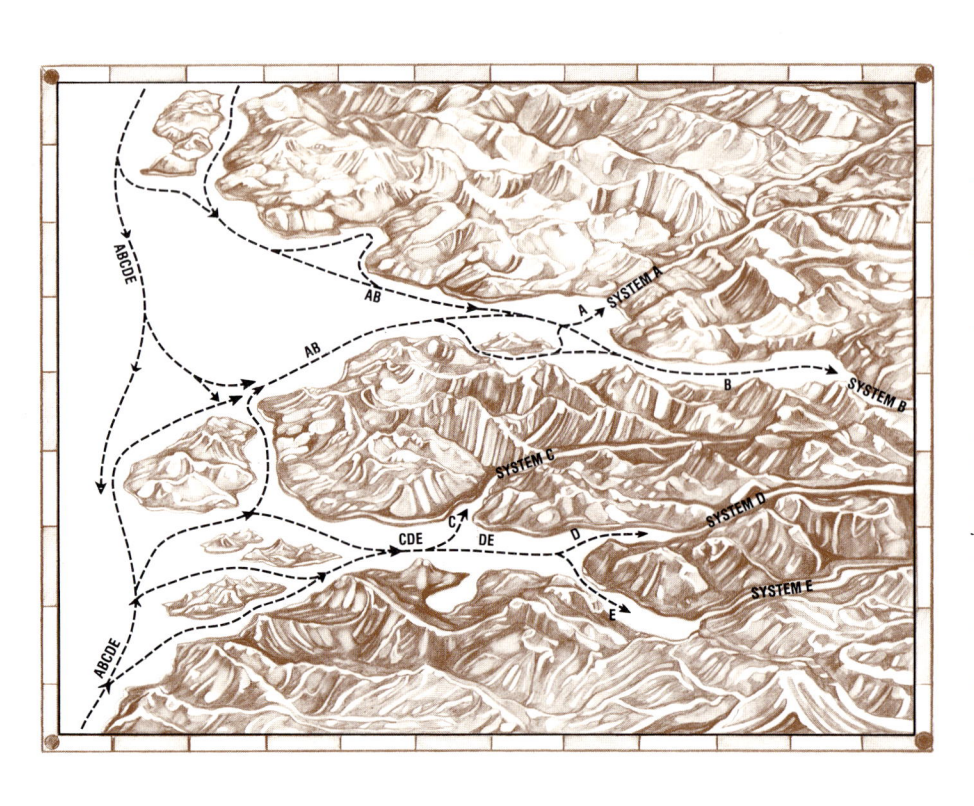

Inshore Fishery Management

TOP MAP
Hypothetical migration routes and timing, in mean days of travel, as an example of how a one-day-per-week fishery in the area outside the fishing boundaries could exploit more than one seventh of the fish available each week.

LOWER MAP
Hypothetical migration routes of spawning stocks (system A, B, C, D, E) as an example of how fisheries in some areas could simultaneously exploit a number of salmon stocks bound for different spawning rivers.

Homing sockeye jumping ten-foot falls

Sockeye holding along banks of Adams River during final maturation

Males holding in groups become restless . . .
. . . prior to entering nearby spawning areas

Male and female courting over nest area

Intruding males harass spawning pair

Dominant male defends nest area against intruding male

148

Sockeye spawning pair

Spawning pair over nest

Persistent male intruder bites nesting female

Female digging nest

The heavy Adams River sockeye run every fourth year attracts thousands of spectators

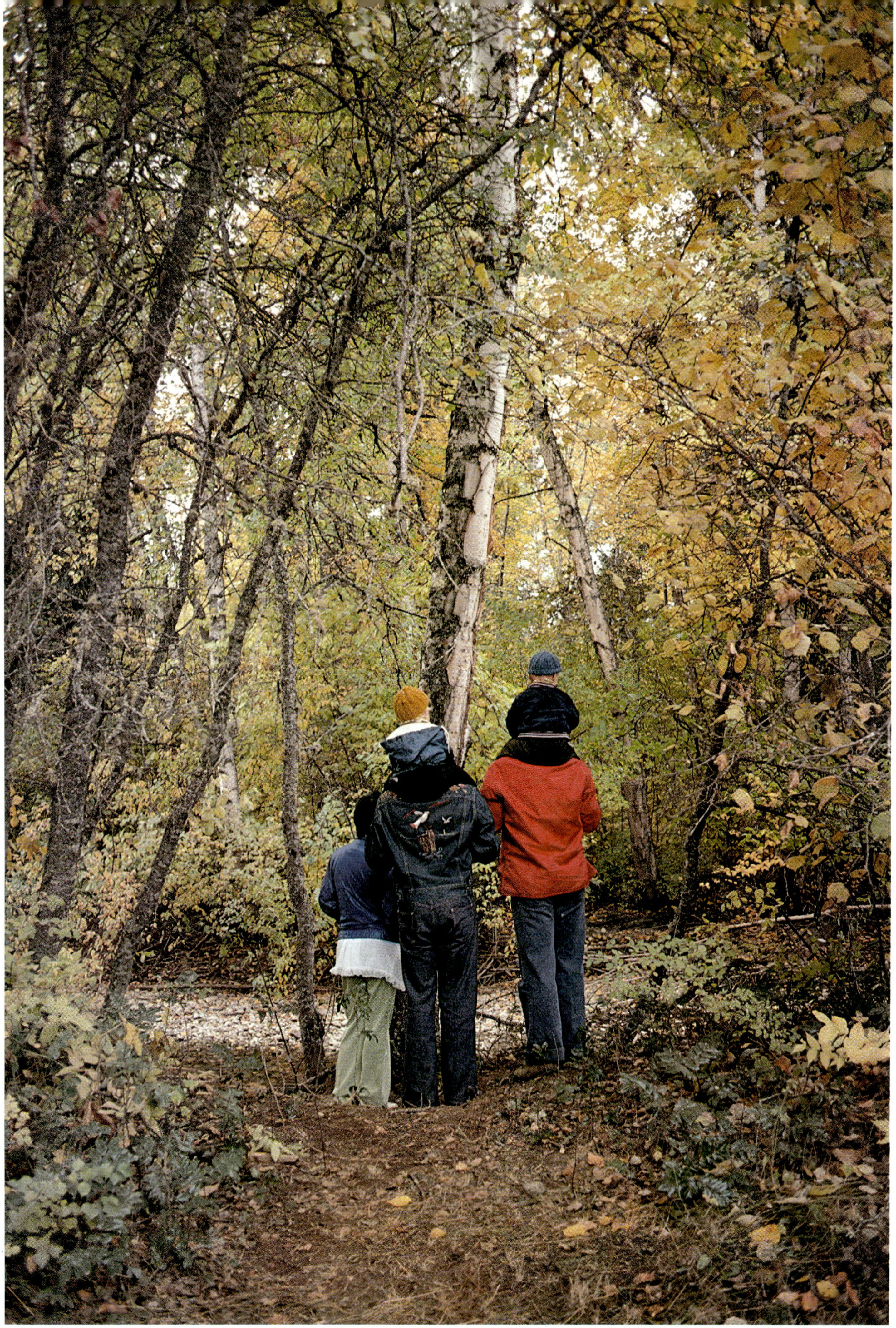

The physical deterioration that causes death is not always evident

Spawned out male sockeye

All Pacific salmon die after spawning

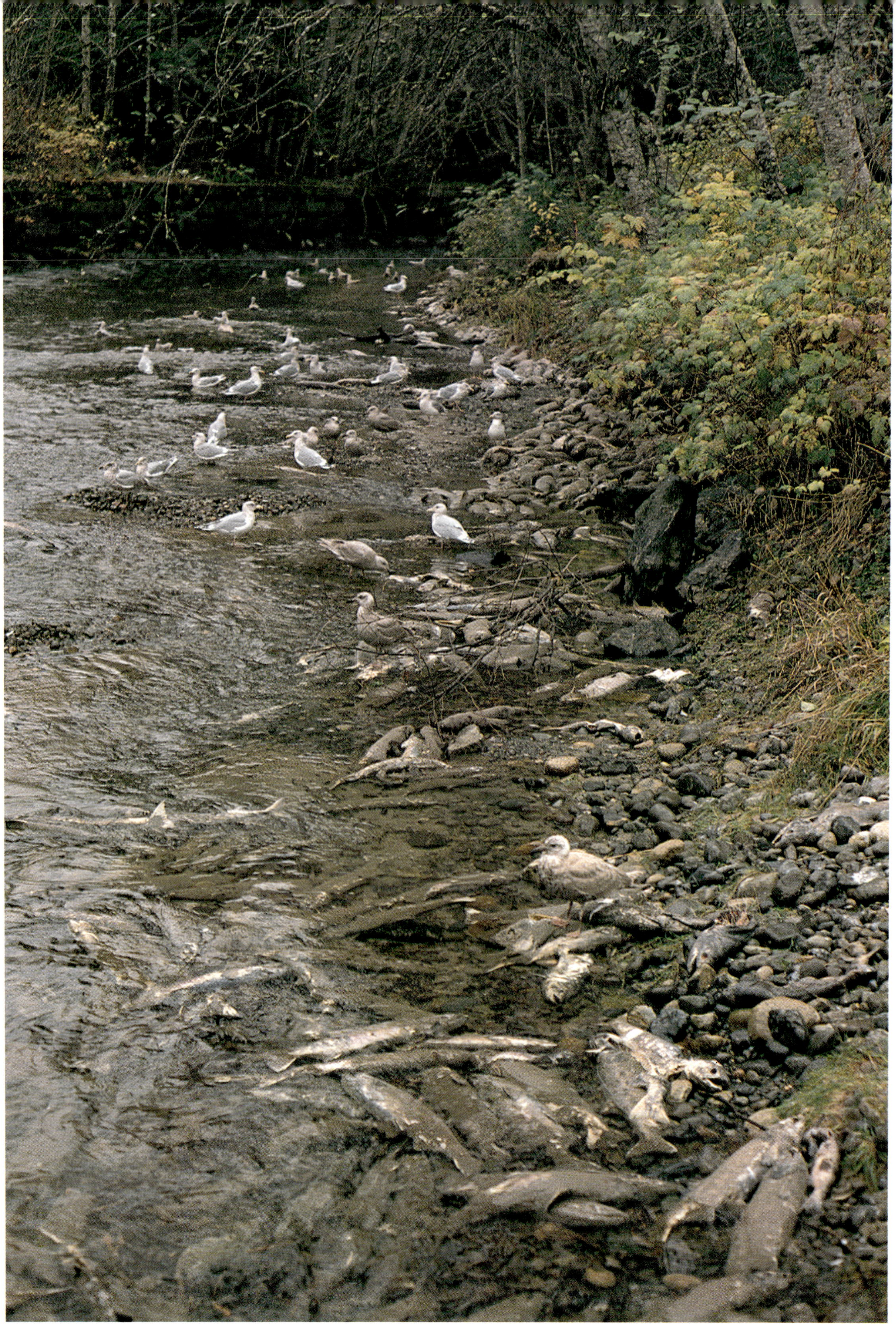

Afterword

Science and technology are pillars on which our civilization rests. Until recently we aspired to claim mastery over nature. Now we are suddenly aware that our massive meddling with natural processes—through agriculture, dams, pesticides—produce results far different from, and more serious than, those anticipated.

The program to increase salmon populations in British Columbia is one such intrusion. The state of Alaska is proposing an even larger salmon enhancement program, and Washington and Oregon states may undertake a series of new salmon hatcheries to offset declining catches. Such increases in salmon numbers may only result in more severe competition for food in the open ocean and destruction of wild stocks. The big financial rewards hoped for by each of these governments may for reasons unforeseen not materialize.

Today, people are awakening to the uncertainties that result when man interferes with complex ecological systems. We pride ourselves that we are living at a time of ecological awareness. But have we really progressed?

More than a hundred years ago, Chief Seathl of the Duwamish tribe in the state of Washington, for whom the largest city of that state is named, dictated a letter to the president of the United States. The letter—written in 1855—concerned the proposal to purchase the tribe's land.

The Great Chief in Washington sends word that he wishes to buy our land. The Great Chief also sends us words of friendship and goodwill. This is kind of him, since we know he has little need of our friendship in return. But we will consider your offer, for we know if we do not do so, the white man may come with guns and take our land. What Chief Seathl says, the Great Chief in Washington can count on as truly as our white brothers can count on the return of the season. My words are like the stars—they do not set.

How can you buy or sell the sky—the warmth of the land? The idea is strange to us. We do not own the freshness of the air or the sparkle of the water. How can you buy them from us? We will decide in our time. Every part of this earth is sacred to my people. Every shining pine needle, every sandy shore, every mist in the dark woods, every clearing and humming insect is holy in the memory and experience of my people.

We know that the white man does not understand our ways. One portion of the land is the same to him as the next, for he is a stranger who comes in the night and takes from the land whatever he needs. The earth is not his brother, but his enemy, and when he has conquered it, he moves on. He leaves his fathers' graves behind and does not care. He kidnaps the earth from his children. He does not care.

Our children have seen their fathers humbled in defeat. Our warriors have felt shame. And after defeat, they turn their days in idleness and contaminate their bodies with sweet food and strong drink. It matters little where we pass the rest of our days—they are not many. A few more hours, a few more winters, and none of the children of the great tribes that once lived on the earth, or that roamed in small bands in the woods, will be left to mourn the graves of a people once as powerful and hopeful as yours.

One thing we know which the white man may one day discover. Our God is the same God. You may think now that you own our land. But you cannot. He is the God of man. And His compassion is equal for the red man and the white. The earth is precious to Him. And to harm the earth is to heap contempt on its creator.

The whites, too, shall pass—perhaps sooner than other tribes. Continue to contaminate your bed, and you will one night suffocate in your own waste. When the buffalo are all slaughtered, the wild horses all tamed, the secret corners of the forest heavy with the scent of many men and the view of the ripe hills blotted by talking wives, where is the thicket? Gone. Where is the eagle? Gone. And what is it to say goodbye to the swift and the hunt, [it is] the end of living and the beginning of survival.

We might understand if we knew what it was that the white man dreams, what hopes he describes to his children on long winter nights, what visions he burns into their minds, so that they will wish for tomorrow. But we are savages. The white man's dreams are hidden from us. And because they are hidden, we will go our own way. If we agree, it will be to secure the reservation you have promised. There perhaps we may live out our brief days as we wish.

When the last red man has vanished from the earth, and the memory is only the shadow of a cloud moving across the prairie, these shores and forest will still hold the spirits of my people, for they love this earth as the newborn loves its mother's heartbeat. If we sell you our land, love it as we've loved it. Care for it, as we've cared for it. And with all your strength, with all your might, and with all your heart—preserve it for your children, and love it as God loves us all. One thing we know—our God is the same God. The earth is precious to Him. Even the white man cannot be exempt from the common destiny.

His fathers' graves and his children's birthright are forgotten. His appetite will devour the earth and leave only behind a desert. The sight of your cities pains the eyes of the red man. But perhaps it is because the red man is a savage and does not understand. . .

There is no quiet place in the white man's cities. No place to hear the leaves of spring or the rustle of insects' wings. But perhaps because I am a savage and do not understand—the clatter only seems to insult the ears. And what is there to life if a man cannot hear the lovely cry of a whippoorwill or the arguments of the frogs around a pond at night? The Indian prefers the soft sound of the wind darting over the face of the pond, and the smell of the wind itself cleansed by a midday rain, or scented with pinion pine. The air is precious to the red man. For all things share the same breath—the beasts, the trees, the man. The white man does not seem to notice the air he breathes. Like a man dying many days, he is numb to the stench.

If I decide to accept, I will make one condition. The white man must treat the beasts of this land as his brothers. I am a savage and I do not understand any other way. I have seen a thousand rotting buffalo on the prairies, left by the white man who shot them from a passing train. I am a savage and I do not understand how the smoking iron horse can be more important than the buffalo that we kill only to stay alive. What is man without the beasts? If all the beasts were gone, man would die from great loneliness of spirit, for whatever happens to the beast also happens to the man. All things are connected. Whatever befalls the earth befalls the sons of the earth.